AF208102

Laboratory Explorations to Accompany

Microelectronic Circuits

Eighth Edition
By Adel S. Sedra, Kenneth C. Smith,
Tony Chan Carusone, and Vincent Gaudet

Vincent C. Gaudet
University of Waterloo

Kenneth C. Smith
University of Toronto

New York   Oxford
OXFORD UNIVERSITY PRESS

Oxford University Press is a department of the University of Oxford.
It furthers the University's objective of excellence in research, scholarship,
and education by publishing worldwide.

Oxford   New York
Auckland   Cape Town   Dar es Salaam   Hong Kong   Karachi
Kuala Lumpur   Madrid   Melbourne   Mexico City   Nairobi
New Delhi   Shanghai   Taipei   Toronto

With offices in
Argentina   Austria   Brazil   Chile   Czech Republic   France   Greece
Guatemala   Hungary   Italy   Japan   Poland   Portugal   Singapore
South Korea   Switzerland   Thailand   Turkey   Ukraine   Vietnam

Copyright © 2020, 2015, 2014, 2011, 2008 by Oxford University Press

For titles covered by Section 112 of the US Higher Education Opportunity
Act, please visit www.oup.com/us/he for the latest information about
pricing and alternate formats.

Published by Oxford University Press
198 Madison Avenue, New York, NY 10016
www.oup.com

Oxford is a registered trademark of Oxford University Press.

All rights reserved. No part of this publication may be reproduced,
stored in a retrieval system, or transmitted, in any form or by any means,
electronic, mechanical, photocopying, recording, or otherwise,
without the prior permission of Oxford University Press.

ISBN: 978-0-19-750857-2

Printed in Canada
on acid-free paper

List of Experiments

All references are to *Microelectronic Circuits,* Eighth Edition, by Sedra, Smith, Chan Carusone and Gaudet

PREFACE

OUTLINE OF THIS MANUAL

Laboratory Explorations accompanies the text *Microelectronic Circuits,* Eighth Edition, by Adel S. Sedra, Kenneth C. Smith, Tony Chan Carusone, and Vincent Gaudet and includes an extensive list of laboratory experiments to complement Part I of the textbook. Each experiment can be run in a conventional microelectronic circuits laboratory setting or may use an instrumentation board to assist undergraduate students in their electronics education.

A course that covers Part I of the textbook usually assumes a prerequisite course in linear circuit analysis, as well as familiarity with differential and integral calculus. Some students will have had exposure to differential equations, but few will be familiar with frequency-domain analysis. Device concepts and semiconductor physics are often taught in a separate course. The microelectronics course provides students with their first exposure to nonlinear circuits (diode and single-transistor circuits). It is also one of the first courses in an electrical/computer/electronics curriculum that contains a strong *design* component.

The goals of these laboratory experiments are to develop practical skills and to reinforce topics that are covered in the course; the experiments have been chosen to complement the lecture material. Most experiments have some component of design, e.g., where a student has to choose resistor values to meet some design specification. Although specifications are stated as absolutes, we encourage instructors to remind their students that, in microelectronic circuit design, the real expectation is to meet a specification within a certain tolerance. In some labs, guidance has been given on how to select component values. However, instructors should remember that there are many different ways of meeting design specifications, and a bit of creativity can lead to lifelong learning!

The suggested experimental setup in the laboratory includes an oscilloscope, a function generator, a power supply, and a digital multimeter. At the authors' home universities, lab groups sign out a kit that includes a breadboard for prototyping, several 1% tolerance resistors and capacitors (standard values include 100 Ω, 1 kΩ, 10 kΩ, and 100 kΩ), a set of junction diodes, a quad 741 op-amp (LM348), and a few MOSFET and BJT transistors. It is amazing how much can be done with so little! Students are required to prototype the main circuit to be tested, on their breadboard, prior to the lab session.

Ultimately, the goal is for this manual to serve as a complement to the main textbook; as such, much of the expository material has been omitted here, and references are made to specific sections of the book.

Microelectronics is an exciting and constantly changing field. We hope that students will enjoy this material!

LABORATORY STRUCTURE

The laboratory experiments in this manual have been selected to complement the material covered in Part I of the text. Experiments are selected from among the circuits presented in the textbook and the end-of-chapter problem sets.

We begin with six labs (denoted Labs 2.1 to 2.6) on op-amp–based circuits (S&S Chapter 2). Diodes are covered in Labs 4.1 to 4.3 (S&S Chapter 4). MOS-based circuits at DC are covered in Labs 5.1 to 5.4 (S&S Chapter 5), and bipolar transistors at DC are covered in Labs 6.1 to 6.4 (S&S Chapter 6). MOSFET and bipolar transistor amplifiers are covered in Labs 7.1 to 7.16 (S&S Chapter 7). Finally, to make things interesting, Lab 7.17 makes interchangeable use of a MOS and a bipolar transistor in an amplifier circuit.

The experiments in this manual have varying levels of difficulty. We have designed most of them so that they can be completed in approximately 3 hours. This would typically be broken up as follows: (1) 45 minutes of circuit design and simulation, (2) 30 minutes for prototyping, (3) 60 minutes to run the experiment and record results in the student's lab book, and (4) 45 minutes for post-experiment analysis and simulations. All experiments in this manual have been tested by an undergraduate student, Pral Bhojak, for duration and ease of understanding.

Next, we suggest several potential sequences of experiments, accommodating courses with 6 labs and 12 labs. Each sequence begins with a few op-amp and diode-based labs and is followed by several transistor-based labs. We propose sequences that are heavy on MOSFET experiments, others that are heavy on BJT experiments, and finally sequences that include both MOSFET and BJT experiments.

12-lab sequence with a focus on MOSFETs:

- 2.1, 2.4, 2.5, 4.1, 4.2, 4.3, 5.1, 5.2, 5.3, 7.1, 7.3, 7.7

12-lab sequence with a focus on BJTs:

- 2.1, 2.4, 2.5, 4.1, 4.2, 4.3, 6.1, 6.2, 6.3, 7.9, 7.11, 7.15

12-lab sequence with a mixed focus on both MOSFETs and BJTs:

- 2.1, 2.4, 2.5, 4.1, 4.2, 4.3, 5.1, 7.1, 7.7, 6.1, 7.9, 7.17

6-lab sequence with a focus on MOSFETs:

- 2.1, 4.2, 5.1, 5.3, 7.1, 7.7

6-lab sequence with a focus on BJTs:

- 2.1, 4.2, 6.1, 6.3, 7.9, 7.15

6-lab sequence with a mixed focus on both MOSFETs and BJTs:

- 2.1, 4.2, 5.1, 7.1, 6.1, 7.9

MYDAQ VS. IN-LAB MEASUREMENT

National Instruments' myDAQ system includes ±15-V power supplies, analog I/O (analog signal generation and an oscilloscope, both running at 200 kS/s), digital I/O, and a digital multimeter, all of which can be interfaced to a student's own breadboard. In other words, it encompasses most if not all the equipment that would be used in a lab for a typical first course in low-frequency microelectronics, e.g., one that covers Part I of the textbook; more advanced topics, such as frequency response of devices, require more sophisticated lab equipment that runs at higher frequencies, and these are therefore not the focus of the present manual. myDAQ connects to a computer via a USB connection and is controlled by software. Each experiment in this manual can be completed with a computer, a myDAQ board, and a lab kit similar to the one just described, i.e., containing a relatively restricted set of components.

The introduction of myDAQ begs the following question: Should we change our approach to microelectronics pedagogy? Let us first look at several ways in which myDAQ could be incorporated into a course.

Using myDAQ as a complement to regularly scheduled laboratory sessions

In this model, students are encouraged to prototype the circuit for an experiment in advance of a lab session. With myDAQ, they can verify the circuit's functionality and take some preliminary measurements, e.g., DC bias points. Under these circumstances, the lab session may be run quite differently than it is now. Lab instructors and teaching assistants can become more effective by focusing on concepts and on experimentation (and exploration!) rather than on figuring out how a circuit has been incorrectly wired. It also helps with the lab availability issue, since most of the "tedious" lab work can now be done outside normal lab hours.

Using myDAQ as a replacement for laboratory sessions

In smaller universities and colleges, running a fully equipped laboratory facility can be an expensive proposition. Given the low cost of purchasing an entire myDAQ-based system, microelectronics courses can now be introduced to universities and colleges where no such course existed before, especially if the cost of the board is amortized over several courses (e.g., the prerequisite linear circuits course, a course based on the textbook, and a design project course).

Using myDAQ in supplementary experiments to complement lectures/tutorials

With myDAQ, it becomes reasonable to expect students to prototype a circuit under study, e.g., from a problem set, during their self-study time. Much of the cabling normally associated with lab experiments in microelectronics is integrated onto myDAQ, and hence the prototyping time is shortened. After design and simulation using MultiSim, a circuit from a problem assignment can be rapidly prototyped and measured in a matter of minutes. Hence, it may make sense for an instructor to assign myDAQ-based exercises in class or as part of a problem set. We suggest using myDAQ for a few tutorials.

STUDENT LOGBOOK STRUCTURE

There is debate among faculty and lab instructors about what students should write in their logbooks. Should lab reports be long and descriptive or short and tabular? Should students describe their methodology or focus on recording measurement results? Should lab reports be written up during the lab session or afterwards? Which results should students enter into their logbook? Should we quiz students on their lab results? How can instructors provide useful guidance and feedback?

There may be no single best answer to the foregoing questions, and the principle of academic freedom demands that all instructors use their best judgment based on student background preparation, personal preference, and local considerations. But we will at least offer our own opinion: Less is More!

The scientific method has a tendency to work well when hypotheses are kept simple: Study the effect of one variable at a time! Experiments in microelectronics are no different. Study one circuit. Examine one principle. For this reason the experiments in this manual have been kept short and to the point, and they refer to specific principles that are cross-referenced in the main text. Students should focus on recording key results that relate directly to that principle: What were the operating conditions (e.g., voltages at the DC operating point on a clearly annotated circuit diagram), key waveforms (e.g., input and output signals), and important measurements that relate directly to the circuit (e.g., droop voltage in a peak rectifier, or gain and −3dB point in an amplifier).

To allow experimental concepts to sink in, it may also be useful to quiz students on what they learned in an experiment. There are many opportunities during a term: during lab sessions, tutorials, midterms, and final exams. Potential questions can be drawn from the "Post-Measurement Exercises" section in each lab. This type of testing reinforces the notion that experimentation is a critical part of the learning process that does not end with the completion of the experiment and handing in of a lab report for grading, but includes an understanding of what was actually done.

Finally, we encourage all students to include a short concluding statement in all labs, indicating the main points that were learned. As educators, we like to think that our students are learning to think; asking for such concluding statements provides each student an opportunity to pause and reflect on what has been learned.

DEVICE CHARACTERIZATION

We recommend that students do labs 5.1 (NMOS) and 5.2 (PMOS) prior to the other MOSFET labs and labs 6.1 (NPN) and 6.2 (PNP) prior to the other BJT labs. In these labs, students characterize their transistors and extract parameters that can be used in further hand analysis and simulations. Based on measurements, we have extracted the following parameters for some of the diodes and transistors used in our own experiments:

1N4003 diode: $i_D = 0.315$ mA @ $v_D = 0.70$ V
MC14007 NMOS: $k_n = 1.08$ mA/V^2, $V_{tn} = 1.45$ V, $V_A = -68.5$ V
MC14007 PMOS: $k_p = 1.56$ mA/V^2, $V_{tp} = -1.5$ V, $V_A = 12.1$ V
NTE2321 NPN: $\beta = 18.46$, $V_A = -16.2$ V
NTE2322 PNP: $\beta = 106.8$, $V_A = 4.82$ V

Datasheets for these devices are all easily available online. Instructors may wish to collect some at the beginning of the term to hand them out to the class. Several experiments require large coupling capacitors. Usually, you will need electrolytic capacitors that require appropriate polarization. Other experiments require a current measurement using a power supply's internal digital current meter. If this is not available, you may need to indirectly measure current through a voltage measurement across a small series resistor.

KEY FEATURES

- Experiments are written in a concise way, with clear steps.
- Coverage includes both MOSFET and bipolar devices, including PMOS, NMOS, NPN, and PNP transistors.
- Experiments start from concepts and hand analysis and include simulation, measurement, and post-measurement discussion components.
- Experiments are designed to be completed either in a traditional laboratory setting or using on instrumentation board such as the National Instruments myDAQ board.

ACKNOWLEDGMENTS

The authors wish to acknowledge the many people who have helped us along the way. In particular, faculty colleagues (in particular, David Nairn, Siddharth Garg), graduate students (Brendan Crowley, Karl Jensen), and laboratory instructors (Manisha Shah, Paul Hayes) have provided several valuable insights. We also thank Dan Sayre and Micheline Frederick from Oxford University Press for making it all possible. Several anonymous external reviewers provided detailed comments that have made this a much better laboratory manual than would otherwise have been possible. A big thank you goes to Pral Bhojak, a former undergraduate student at the University of Waterloo, who patiently went through each experiment, sometimes many times over with new specifications, and gave us guidance from the student's perspective.

Finally, we dedicate this manual to Katherine, Margaret, and Laura, who make it all worthwhile!

Inverting Op-Amp Configuration
(See Section 2.2, p. 64 of Sedra/Smith)

OBJECTIVES:

To study an operational amplifier and an inverting amplifier by:

- Completing the analysis of the circuit and selecting resistors that satisfy design specifications for two values of voltage gain.
- Simulating the circuits to compare the results with the paper analysis.
- Implementing the circuits in an experimental setting, taking measurements, and comparing circuit performance to theoretical and simulated results.

MATERIALS:

- Laboratory setup, including breadboard
- 1 741-type operational amplifier (obtain its datasheet)
- Several wires and resistors of varying sizes

PART 1: DESIGN AND ANALYSIS

Consider the circuit shown in Figure L2.1:

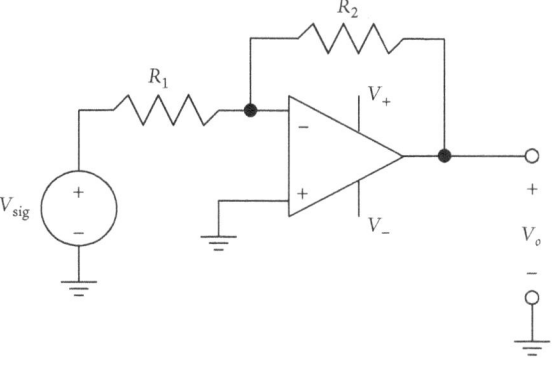

FIGURE L2.1: Inverting amplifier circuit. See Fig. 2.5, p. 64 S&S.

Design two versions of the circuit in Figure L2.1: one that achieves $A_v = -10$ V/V, and one that achieves $A_v = -100$ V/V. Assume a 1-kHz input waveform v_{sig} of

50 mV$_{\text{pk–pk}}$. Your circuit may only draw up to 50 μA of current from v_{sig} (it will draw more from V_+ and V_-). Use supplies of $V_+ = -V_- = 15$ V.

Hand calculations

- Sketch the circuit in your lab book, clearly labeling the op-amp terminals.
- What values of R_1 must you use to satisfy the current constraint? What is the input resistance of the circuit?
- What values of R_2 do you need to use to meet the two gain specifications? Is the problem completely specified?

Simulation

- Simulate both circuits using a transient simulation with a 50-mV$_{\text{pk–pk}}$ 1-kHz input sine wave v_{sig}. In your simulations, assume that your input voltage source v_{sig} has an output resistance of 50 Ω. Use values of R_1 and R_2 based on your preceding calculations.
- Plot the input and output waveforms. What is the DC voltage at the inverting terminal of the op-amp?
- What are the simulated voltage gains of your circuit?

PART 2: PROTOTYPING AND MEASUREMENT

- Assemble the circuit with $A_v = -10$ V/V onto a breadboard. Do not include the 50-Ω output resistance of v_{sig}.
- Using a digital multimeter, measure the DC voltages at the input, output, and inverting terminal, while leaving the input tied to ground.
- Using a function generator, provide a 1-kHz 50-mV$_{\text{pk–pk}}$ sine wave to the input. Using an oscilloscope, observe the output voltage waveform.
- Using a power supply, provide a DC input to the circuit in increments of 0.1 V, from –1 V to +1 V. Record the values of v_O and plot your results.
- Repeat the measurements for the gain $A_v = -100$ V/V, but this time sweeping the DC input to the circuit in increments of 0.01 V, from –0.1 V to +0.1 V.
- Using a digital multimeter, measure all resistors to three significant digits.

PART 3: POST-MEASUREMENT EXERCISE

- For both circuits, calculate the voltage gains you obtained in measurement. Explain any discrepancies between the experiments, simulations, and hand analysis.
- Recalculate the theoretical gains of the circuit using the measured resistor values. Are the recalculated values closer to your measured gains?

PART 4 [OPTIONAL]: EXTRA EXPLORATION

- In your measurement setup, gradually increase the frequency of the input sine wave until the output's amplitude is about 70% of what it was at lower frequencies. At what frequency does this happen? This represents the –3-dB frequency of the circuit.
- In your measurement setup, gradually increase the amplitude of the input sine wave until the output becomes distorted. At what amplitude does this begin to happen? Can you explain this phenomenon?

Non-Inverting Op-Amp Configuration
(See Section 2.3, p. 74 of Sedra/Smith)

OBJECTIVES:

To study an operational amplifier and a non-inverting amplifier by:

- Completing the analysis of the circuit and selecting resistors that satisfy design specifications for voltage gain.
- Simulating the circuits to compare the results with the paper analysis.
- Implementing the circuits in an experimental setting, taking measurements, and comparing performance with theoretical and simulated results.

MATERIALS:

- Laboratory setup, including breadboard
- 1 741-type operational amplifier (obtain its datasheet)
- Several wires and resistors of varying sizes

PART 1: DESIGN AND ANALYSIS

Consider the circuit shown in Figure L2.2:

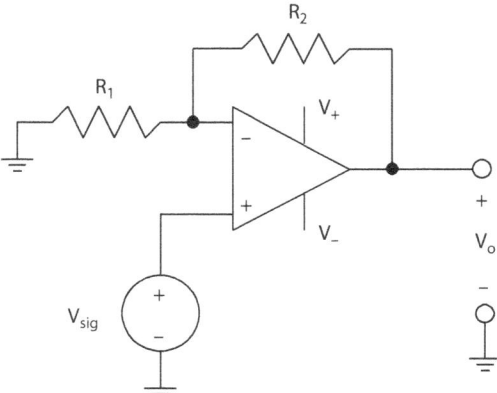

FIGURE L2.2: Non-inverting amplifier circuit. See Fig. 2.12, p. 75 S&S.

Design the circuit in Figure L2.2 such that $A_v = 11$ V/V. Assume an input waveform v_{sig} of 50 mV$_{pk-pk}$. Use supplies of $V_+ = -V_- = 15$ V.

Hand calculations

- Sketch the circuit in your lab book, clearly labeling the op-amp terminals.
- What values of R_1 and R_2 do you need to use to meet the gain specification? Is the problem completely specified? If not, what needs to be specified?

Simulation

- Simulate your circuit using a 50-mV$_{pk-pk}$ 1-kHz input sine wave v_{sig}. In your simulation, assume your input voltage source v_{sig} has an output resistance of 50 Ω. Use values of R_1 and R_2 based on your preceding calculations.
- Plot the input and output waveforms. What is the DC voltage at the inverting terminal of the op-amp?
- What is the gain of your circuit?

PART 2: PROTOTYPING AND MEASUREMENT

- Assemble the circuit onto a breadboard. Do not include the 50-Ω output resistance of v_{sig}.
- Using a digital multimeter, measure the DC voltages at the input, output, and inverting terminal, while leaving the input tied to ground.
- Using a function generator, provide a 1-kHz 50-mV$_{pk-pk}$ sine wave to the input. Using an oscilloscope, observe the output voltage waveform.
- Using a power supply, provide a DC input to the circuit in increments of 0.1 V, from –1 V to 1 V. Record the values of v_O and plot your results.
- Using a digital multimeter, measure all resistors to three significant digits.

PART 3: POST-MEASUREMENT EXERCISE

- Calculate the voltage gain you obtained in your measurement. Explain any discrepancies between the experiments, simulations, and hand analysis.
- Recalculate the theoretical gain of the circuit using the measured resistor values. Is the recalculated value closer to the measured gain?

PART 4 [OPTIONAL]: EXTRA EXPLORATION

- In your measurement setup, gradually increase the frequency of the input sine wave until the output amplitude is about 70% of what it was at lower frequencies. At what frequency does this happen? This represents the –3-dB frequency of the circuit.
- In your measurement setup, gradually increase the amplitude of the input sine wave until the output becomes distorted. At what amplitude does this begin to happen? Can you explain this phenomenon?

Difference Amplifier
(See Section 2.4, p. 78 of Sedra/Smith)

OBJECTIVES:

To study an operational amplifier and a difference amplifier circuit by:

- Completing the analysis of the circuit, and selecting resistors that satisfy design specifications.
- Simulating the circuit to compare the results with the paper analysis.
- Implementing the circuit in an experimental setting, taking measurements, and comparing its performance with theoretical and simulated results.

MATERIALS:

- Laboratory setup, including breadboard
- 1 741-type operational amplifier (obtain its datasheet)
- Several wires and resistors of varying sizes

PART 1: DESIGN AND ANALYSIS

Consider the circuit shown in Figure L2.3:

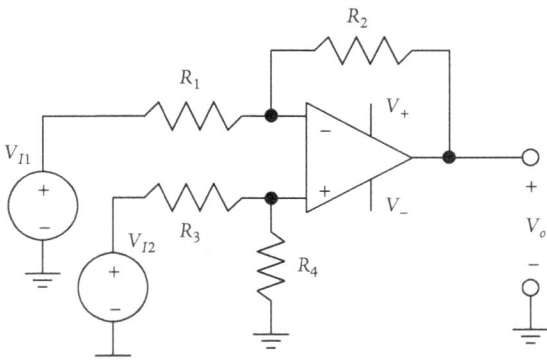

FIGURE L2.3: Difference amplifier circuit. See Fig. 2.16, p. 80 S&S.

Design the circuit in Figure L2.3 such that $A_d = 10$ V/V and $A_{cm} = 0$. Select resistor values such that the differential input resistance $R_{id} = R_1 + R_3 = 2$ kΩ. Assume input waveforms v_{I1} and v_{I2} of 50 mV$_{pk-pk}$. Use supplies of $V_+ = -V_- = 15$ V.

Hand calculations

- Sketch the circuit in your lab book, clearly labeling the op-amp terminals.
- What values of R_1, R_2, R_3, and R_4 do you need to use to meet the gain and input resistance specifications? Is the problem completely specified?

Simulation

- Use a 50-mV$_{pk-pk}$ 1-kHz input sine wave applied to v_{I1} and another 50-mV$_{pk-pk}$ 1-kHz input sine wave applied to v_{I2} that is 180° out of phase with v_{I1}. In your simulation, assume your input voltage sources have an output resistance of 50 Ω. What are V_{Id} and V_{Icm}? What are V_{Od} and V_{Ocm}?
- Plot the input and output waveforms.
- Report the DC voltage at the inverting terminal and the output of the op-amp.
- What is the simulated value of the differential gain?

PART 2: PROTOTYPING AND MEASUREMENT

- Assemble the circuit onto a breadboard. Do not include the 50-Ω output resistance of your signal sources.
- Using a digital multimeter, measure the DC voltages at each terminal. Leave both inputs grounded.
- While leaving v_{I2} grounded, provide a DC input to v_{I1} in increments of 0.2 V, from –2 V to 2 V. Record the values of v_O and plot your results.
- While leaving v_{I1} grounded, provide a DC input to v_{I2} in increments of 0.2 V, from –2 V to 2 V. Record the values of v_O and plot your results.
- Using a function generator, provide a 1-kHz 50-mV$_{pk-pk}$ sine wave to input v_{I1} and ground input v_{I2}. Using an oscilloscope, capture the output voltage waveform.
- Using a function generator, provide a 1-kHz 50-mV$_{pk-pk}$ sine wave to input v_{I2} and ground input v_{I1}. Using an oscilloscope, capture the output voltage waveform.
- Using a digital multimeter, measure all resistors to three significant digits.

PART 3: POST-MEASUREMENT EXERCISE

- Calculate the values of A_d and A_{cm} obtained in your measurement. What is the common-mode rejection ratio (CMRR) of the circuit? Express the CMRR in units of decibels. Explain any discrepancies between the experiments, simulations, and hand analysis.
- Recalculate the theoretical gains of the circuit, using the measured resistor values. Are the recalculated values closer to your measured gains?

PART 4 [OPTIONAL]: EXTRA EXPLORATION

- In your measurement setup, replace R_2 with a resistor that is 10% smaller in value and remeasure A_d and A_{cm}. How do their values change? What do you conclude?

Instrumentation Amplifier
(See Section 2.4.2, p. 83 of Sedra/Smith)

OBJECTIVES:

To study an instrumentation amplifier circuit by:

- Completing the analysis of the circuit and selecting resistors that satisfy design specifications.
- Simulating the circuit to compare the results with the paper analysis.
- Implementing the circuit in an experimental setting, taking measurements, and comparing its performance with theoretical and simulated results.

MATERIALS:

- Laboratory setup, including breadboard
- Three 741-type operational amplifiers
- Several wires and resistors of varying sizes

PART 1: DESIGN AND ANALYSIS

Consider the circuit shown in Figure L2.4:

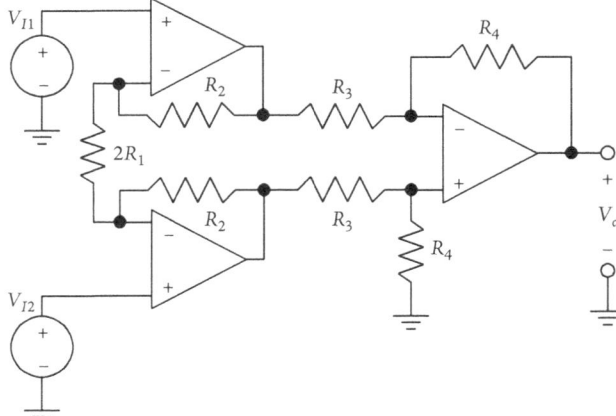

FIGURE L2.4:
Instrumentation amplifier.
Based on Fig. 2.20(b),
p. 84 S&S.

Design the circuit in Figure L2.4 such that A_d = 110 V/V. Select resistor values such that the first stage provides a gain of 11 V/V (magnitude) and R_1 = 1 kΩ and the second stage provides a gain of –10 V/V. Use supplies of V_+ = $-V_-$ = 15 V.

Hand calculations

- Sketch the circuit in your lab book, clearly labeling the op-amp terminals.
- What values of R_1, R_2, R_3, and R_4 do you need to use to meet the gain and input resistance specifications? Is the problem completely specified?

Simulation

- Use a 50-mV$_{pk-pk}$ 1-kHz input sine wave applied to v_{I1} and another 50-mV$_{pk-pk}$ 1-kHz input sine wave applied to v_{I2} that is 180° out of phase with V_{I1}. In your simulation, assume your input voltage sources have an output resistance of 50 Ω. What are V_{Id} and V_{Icm}? What are V_{Od} and V_{Ocm}?
- Plot the input and output waveforms for all simulations.
- For all simulations, report the DC voltage at the inverting terminal and output of each op-amp.
- What are the simulated values of differential and common-mode gain?

PART 2: PROTOTYPING AND MEASUREMENT

- Assemble the circuit onto a breadboard. Do not include the 50-Ω output resistance of your signal sources.
- While leaving v_{i2} grounded, provide a DC input to v_{I1} in increments of 0.01 V, from –0.1 V to +0.1 V. Record the values of v_O and plot your results.
- While leaving v_{i1} grounded, provide a DC input to v_{I2} in increments of 0.01 V, from –0.1 V to +0.1 V. Record the values of v_O and plot your results.
- Using a function generator, provide a 1-kHz 50-mV$_{pk-pk}$ sine wave to input v_{I1} and ground input v_{I2}. Using an oscilloscope, capture the output voltage waveform.
- Using a function generator, provide a 1-kHz 50-mV$_{pk-pk}$ sine wave to input v_{I2} and ground input v_{I1}. Using an oscilloscope, capture the output voltage waveform.
- Using a digital multimeter, measure all resistors to three significant digits.

PART 3: POST-MEASUREMENT EXERCISE

- Calculate the values of A_d and A_{cm} obtained in your measurement. What is the common-mode rejection ratio (CMRR) of the circuit? Express the CMRR in units of decibels. Explain any discrepancies between the experiments, simulations, and hand analysis.

• Recalculate the theoretical gains of the circuit, using the measured resistor values. Are the recalculated values closer to your measured gains?

PART 4 [OPTIONAL]: EXTRA EXPLORATION

• In your measurement setup, replace R_4 with a resistor that is 10% smaller in value and remeasure A_d and A_{cm}. How do their values change? What do you conclude?

Lossy Integrator
(See Section 2.5.1–2.5.2, p. 88 of Sedra/Smith)

OBJECTIVES:

To study a lossy integrator and its time-domain and frequency-domain behavior by:

- Completing the analysis of the circuit and selecting resistors and capacitors that satisfy design specifications.
- Simulating the circuits to compare the results with the paper analysis.
- Implementing the circuit in an experimental setting, taking measurements, and comparing its performance with theoretical and simulated results.

MATERIALS:

- Laboratory setup, including breadboard
- 1 741-type operational amplifier (obtain its datasheet)
- Several wires
- Resistors and capacitors of varying sizes

PART 1: DESIGN AND ANALYSIS

Consider the circuit shown in Figure L2.5:

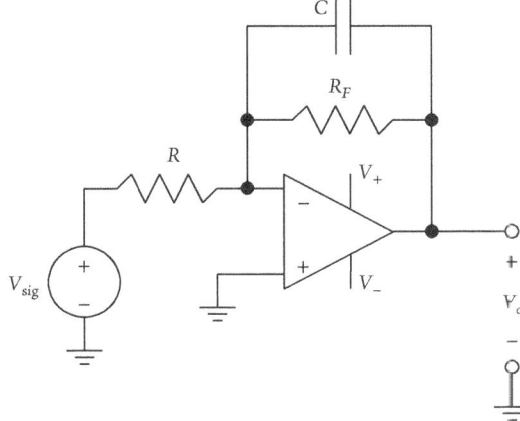

FIGURE L2.5: Lossy integrator.
See Fig. 2.26, p. 93 S&S.

Design the circuit in Figure L2.5 such that its gain at DC is –10 V/V and its –3-dB frequency is 1 kHz. Select resistor values such that the input resistance is 1 kΩ. Use supplies of $V_+ = -V_- = 15$ V.

Hand calculations

- Sketch the circuit in your lab book, clearly labeling the op-amp terminals.
- What values of R, R_F, and C do you need to use to meet the gain, –3-dB frequency, and input resistance specifications? Is the problem completely specified?

Simulation

- *Simulation 1:* Perform a transient simulation of your circuit. Use a 50-mV$_{pk-pk}$ input sine wave. In your simulation, assume your input voltage sources have an output resistance of 50 Ω. Simulate your circuit at frequencies of 100 Hz, 500 Hz, 1 kHz, 2 kHz, and 5 kHz. What do you observe? Plot the input and output waveforms for all simulations.
- *Simulation 2:* Perform an AC simulation of your circuit. In your simulation, assume your input voltage sources have an output resistance of 50 Ω. Plot the magnitude and phase response of your circuit in a Bode plot. Based on this simulation, what is the voltage gain at low frequencies, and what is the –3-dB frequency?

PART 2: PROTOTYPING AND MEASUREMENT

- Assemble the circuit onto a breadboard. Do not include the 50-Ω output resistance of your signal sources.
- Using a digital multimeter, measure the DC voltages at the input, output, and inverting terminal, while leaving the input grounded.
- Using a function generator, provide a 50-mV$_{pk-pk}$ zero-DC sine wave to the input. Using an oscilloscope, capture the output voltage waveform for input frequencies of 10 Hz, 100 Hz, 500 Hz, 1 kHz, 2 kHz, and 5 kHz.
- Vary the input frequency until the output reaches approximately 70% of its low-frequency magnitude. Record this as the –3-dB frequency.
- Using a digital multimeter, measure all resistors to three significant digits.

PART 3: POST-MEASUREMENT EXERCISE

- Calculate the measured voltage gain at low frequencies.
- Recalculate the theoretical gain and –3-dB point of the circuit, using the measured resistor values. Are the recalculated values closer to your measured values?

PART 4 [OPTIONAL]: EXTRA EXPLORATION

- Apply a 10-Hz 1-V$_{pk-pk}$ zero-DC square wave at the input and plot v_O. What is the maximum rate of change in v_O?

Lossy Differentiator
(See Section 2.5.3, p. 95 of Sedra/Smith)

OBJECTIVES:

To study a lossy differentiator and its time-domain and frequency-domain behavior by:

* Completing the analysis of the circuit and selecting resistors and capacitors that satisfy design specifications.
* Simulating the circuits to compare the results with the paper analysis.
* Implementing the circuit in an experimental setting, taking measurements, and comparing its performance with theoretical and simulated results.

MATERIALS:

* Laboratory setup, including breadboard
* 1 741-type operational amplifier (obtain its datasheet)
* Several wires
* Resistors and capacitors of varying sizes

PART 1: DESIGN AND ANALYSIS

Consider the circuit shown in Figure L2.6:

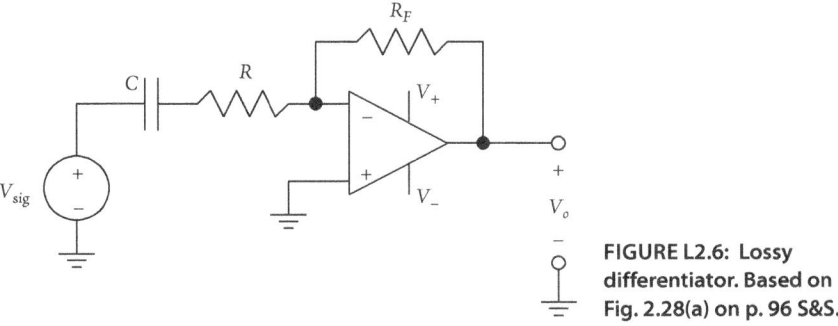

FIGURE L2.6: Lossy differentiator. Based on Fig. 2.28(a) on p. 96 S&S.

Design the circuit in Figure L2.6 such that its high-frequency gain is –10 V/V, and its –3-dB frequency is 100 Hz. Select resistor values such that the input resistance

13

at high frequencies (i.e., where the capacitor acts like a short circuit) is 1 kΩ. Use supplies of $V_+ = -V_- = 15$ V.

Hand calculations

- Sketch the circuit in your lab book, clearly labeling the op-amp terminals.
- What values of R, R_F, and C do you need to use to meet the gain, −3-dB frequency, and input resistance specifications? Is the problem completely specified?

Simulation

- *Simulation 1:* Perform a transient simulation of your circuit. Use a 50-mV$_{pk-pk}$ input sine wave. In your simulation, assume your input voltage sources have an output resistance of 50 Ω. Simulate your circuit at frequencies of 100 Hz, 500 Hz, 1 kHz, 2 kHz, and 5 kHz. What do you observe? Plot the input and output waveforms for all simulations.
- *Simulation 2:* Perform an AC simulation of your circuit. In your simulation, assume your input voltage sources have an output resistance of 50 Ω. Plot the magnitude and phase response of your circuit. Based on this simulation, what is the voltage gain at high frequencies, and what is the −3-dB frequency?

PART 2: PROTOTYPING AND MEASUREMENT

- Assemble the circuit onto a breadboard. Do not include the 50-Ω output resistance of your signal sources.
- Using a digital multimeter, measure the DC voltages at the input, output, and inverting terminal, while leaving the input grounded.
- Using a function generator, provide a 50-mV$_{pk-pk}$ sine wave to the input. Using an oscilloscope, capture the output voltage waveform for input frequencies of 100 Hz, 500 Hz, 1 kHz, 2 kHz, 5 kHz, and 100 kHz.
- Vary the input frequency until the output reaches approximately 70% of its high-frequency magnitude. Record this as the −3-dB frequency.
- Using a digital multimeter, measure all resistors to three significant digits.

PART 3: POST-MEASUREMENT EXERCISE

- Calculate the measured voltage gain at high frequencies.
- Recalculate the theoretical gain and −3-dB point of the circuit, using the measured resistor values. Are the recalculated values closer to your measured values?

PART 4 [OPTIONAL]: EXTRA EXPLORATION

- Apply a 10-Hz 1-V$_{pk-pk}$ zero-DC square wave at the input and plot v_O. What does the output look like? What is the maximum rate of change in v_O?

Diode I-V Transfer Curve
(See Section 4.2, p. 184 of Sedra/Smith)

OBJECTIVES:

To study junction diode terminal characteristics by:

- Analyzing, simulating, and building a diode-based circuit.
- Taking measurements and applying transformations to obtain the diode I-V curve.

MATERIALS:

- Laboratory setup, including breadboard
- One junction diode (e.g., 1N4003)
- Several wires and a resistor

PART 1: SIMULATION

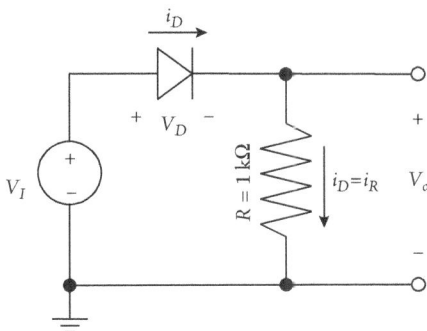

FIGURE L4.1: Circuit used to characterize junction diode terminal characteristics. Based on Fig. 4.21 p. 209 S&S.

Consider the circuit shown in Figure L4.1. Simulate the circuit by varying v_I from –3 V to +3 V in increments of 0.1 V. Generate a plot of i_D vs. v_I and v_O vs. v_I. Do you see a resemblance between the two graphs?

PART 2: MEASUREMENTS

Assemble the circuit onto a breadboard. Using a power supply, vary the input voltage from –3 V to +3 V in increments of 0.25 V. For each point, measure the output voltage v_O using a digital multimeter, and report the current consumption i_D indicated by the power supply. Measure the value of the resistor.

PART 3: POST-MEASUREMENT EXERCISE

- Generate a plot of v_O vs. v_I and a plot of i_D vs. v_I. Since $i_D = v_O/R$, do the two plots generally agree?
- Since the diode voltage is $v_D = v_I - v_O$, generate a new plot of i_D vs. v_D. Is it what you expect?

PART 4 [OPTIONAL]: EXTRA EXPLORATION

- If you have access to a semiconductor parameter analyzer, generate the i_D vs. v_D curve using the analyzer. How does it compare to the curve you generated in Part 3?

Fun with Diodes I: Rectifiers
(See Section 4.6, p. 208 of Sedra/Smith)

OBJECTIVES:

To study diode-based rectifier circuits by:

- Analyzing, simulating, and building several rectifier circuits.
- Noting that many diode-based circuits are easy to assemble. In this lab, you will build several circuits that require only a few simple components.

MATERIALS:

- Laboratory setup, including breadboard
- Several junction diodes (e.g., 1N4003)
- One 741-type operational amplifier
- Several wires, resistors, and capacitors of varying sizes

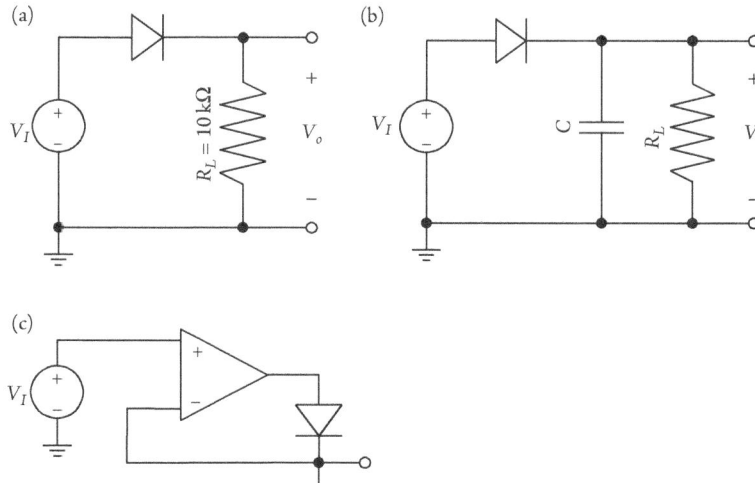

FIGURE L4.2: (a) Half-wave rectifier, (b) peak rectifier, and (c) precision rectifier. Circuits are based on Fig. 4.21 p. 209, Fig. 4.25 p. 216, and Fig. 4.28 p. 222 S&S.

PART 1: SIMULATION

Half-wave rectifier

Consider the half-wave rectifier shown in Figure L4.2(a). Simulate the circuit using a 10-V_{pk-pk} 1-kHz sinusoid and a 1N4003 diode. Provide a plot of v_I and v_O vs. t.

Peak rectifier

Consider the peak detector shown in Figure L4.2(b). Simulate the circuit using a 10-V_{pk-pk} 1-kHz input sinusoid for the two following sets of parameters. For both simulations, provide a plot of v_I and v_O vs. t, and report the peak voltage (V_p) and the ripple voltage (V_r).

- Peak detector I: Use R_L = 1 kΩ, C = 47 μF, 1N4003 diode
- Peak detector II: This time use R_L = 100 Ω, C = 47 μF, 1N4003 diode

Precision rectifier

Consider the precision rectifier shown in Figure L4.2(c). Simulate the circuit using a 10-V_{pk-pk} 1-kHz sinusoidal input, a 741 op-amp, and a 1N4003 diode. Provide a plot of v_I and v_O vs. t. Use R_L = 10 kΩ.

PART 2: MEASUREMENTS

- For each circuit, assemble the circuit, apply the required waveform using a function generator, and capture the input and output voltage waveforms on an oscilloscope.
- For the peak rectifier, record the values of V_p and V_r.
- Using a digital multimeter, measure all resistors to three significant digits.

PART 3: POST-MEASUREMENT EXERCISE

- Using your measured resistor values, resimulate your circuits. How do the updated results compare with your simulations, and experiments? Explain any discrepancies.
- What conclusions do you draw from the two different peak rectifiers?

PART 4 [OPTIONAL]: EXTRA EXPLORATION

Can you turn the precision half-wave rectifier into a precision peak rectifier?

Fun with Diodes II: Limiting and Clamping Circuits
(Various circuits inspired by Chapter 4 of Sedra/Smith)

OBJECTIVES:

To study diode-based limiting and clamping circuits by:

- Analyzing, simulating, and building several circuits, including peak detectors, clamp circuits, and limiter circuits.
- Noting that many diode-based circuits are easy to assemble, in this lab you will build several circuits that require only a few simple components.
- Using an oscilloscope's X-Y mode to plot output vs. input voltage.

MATERIALS:

- Laboratory setup, including breadboard
- Several junction diodes (e.g., 1N4003) and Zener diodes (e.g., 1N4733A)
- Several wires, resistors, and capacitors of varying sizes

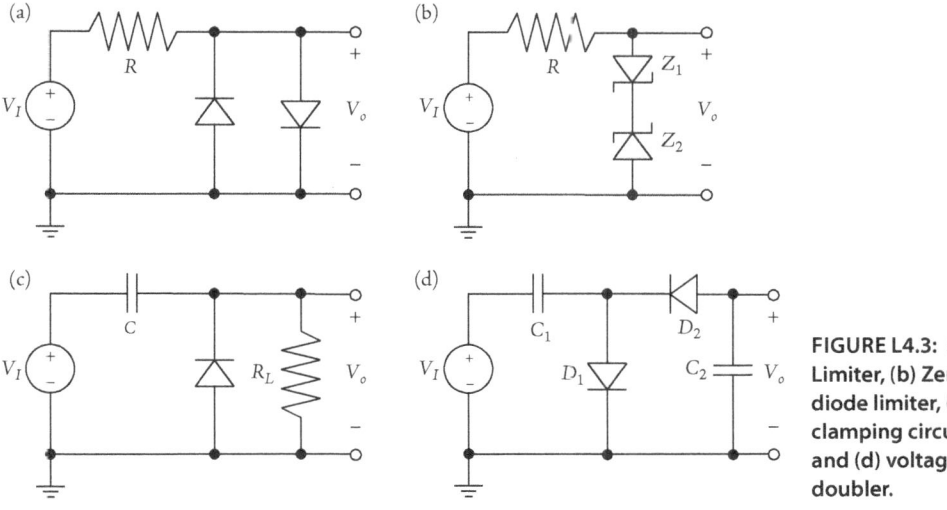

FIGURE L4.3: (a) Limiter, **(b)** Zener diode limiter, **(c)** clamping circuit, and **(d)** voltage doubler.

PART 1: SIMULATION

Consider the circuits shown in Figure L4.3(a)–(d). Simulate each circuit with the parameters indicated next. For each simulation, provide a plot of v_I and v_O vs. t.

- Diode limiter (Figure L4.3(a)):
  - Use $R = 1$ kΩ and 1N4003 diodes.
  - Simulate using a 5-V_{pk-pk} 100-Hz input sinusoid with no DC component.
  - Use your simulator's X-Y mode to plot v_O vs. v_I.
- Zener diode limiter (Figure L4.3(b)):
  - Use $R = 1$ kΩ and 1N4733A Zener diodes.
  - Simulate using a 15-V_{pk-pk} 100-Hz input sinusoid with no DC component.
  - Use your simulator's X-Y mode to plot v_O vs. v_I.
- Clamped capacitor (Figure L4.3(c)):
  - Use $R_L = 10$ kΩ, $C = 47$ μF, and a 1N4003 diode.
  - Simulate using a 2-V_{pk-pk} 100-Hz input square wave with no DC component.
  - What are the highest and lowest voltage values?
- Voltage doubler (Figure L4.3(d)):
  - Use $R_L = 100$ kΩ across the output $C_1 = C_2 = 47$ μF, and 1N4003 diodes.
  - Simulate using a 5-V_{pk-pk} 100-Hz input sinusoid with no DC component.

PART 2: MEASUREMENTS

- For each circuit, build the circuit, apply the input waveform specified above using a function generator, and capture the output voltage waveform on an oscilloscope. For circuits (a)–(c), what are the highest and lowest output voltage values?
- For the limiter circuits, use the oscilloscope's X-Y mode to plot v_O vs. v_I.
- Using a digital multimeter, measure all resistors to three significant digits.
- Further exploration I: Can you change the limiting voltages for the first circuit to approximately +1.4 V and –1.4 V?
- Further exploration II: Can you turn the clamped capacitor into a negative clamp?

PART 3: POST-MEASUREMENT EXERCISE

- Do any of your measurement results differ significantly from what you expect and from the simulations? Explain.

PART 4 [OPTIONAL]: EXTRA EXPLORATION

- Can you modify the voltage doubler so it produces a positive output voltage?

NMOS I-V Characteristics
(See Sections 5.1–5.2, p. 246 of Sedra/Smith)

OBJECTIVES:

To study NMOS transistor I-V curves by:

- Simulating a transistor to investigate the drain current vs. gate-to-source voltage and drain-to-source voltage.
- Implementing a circuit and taking measurements of the I_D vs. V_{GS} and I_D vs. V_{DS} curves.
- Extracting values of k_n, V_{tn}, and λ_n.

MATERIALS:

- Laboratory setup, including breadboard
- 1 enhancement-type NMOS transistor (e.g., MC14007)
- Several wires

PART 1: SIMULATION

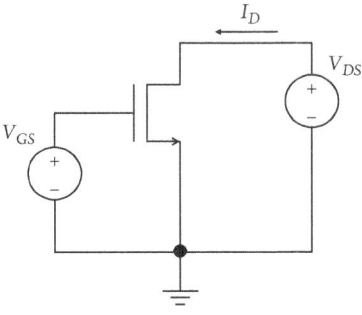

FIGURE L5.1: Transistor measurement circuit. See Table 5.1 in S&S.

Consider the circuit in Figure L5.1. Enter the circuit into your simulator's schematic editor, applying DC voltage supplies to the gate and drain of the transistor.

I_D vs. V_{GS}

While setting V_{DS} to a constant value of 5 V, sweep the gate voltage from 0 V to 5 V in increments of 0.1 V. Plot a curve of I_D vs. V_{GS}. At what value of V_{GS} does the current turn on?

I_D vs. V_{DS}

For three values of V_{GS} (2.5 V, 3.0 V, and 3.5 V), sweep the drain voltage from 0 V to 5 V in increments of 0.1 V. Plot the curves for I_D vs. V_{DS} onto a single graph, clearly indicating the value of V_{GS} next to each curve.

PART 2: MEASUREMENTS

Assemble the circuit from Figure L5.1, using a power supply to generate the DC voltages.

I_D vs. V_{GS}

While setting V_{DS} to a constant value of 5 V, sweep the gate voltage from 1.0 V to 3.5 V in increments of 0.25 V (note, we have reduced the number of data points with respect to the simulations), and measure the drain current using the power supply. (*Note*: Not all power supplies allow you to measure current accurately; if this is the case for your lab setup, you may place a small resistor in series with the drain and measure the voltage drop across the resistor.) Plot a curve of I_D vs. V_{GS}. At what value of V_{GS} does the NMOS turn on?

I_D vs. V_{DS}

For three values of V_{GS} (2.5 V, 3.0 V, and 3.5 V), sweep the drain voltage from 0 V to 3.5 V in increments of 0.5 V, and measure the drain current using the power supply. Plot the curves for I_D vs. V_{DS} onto a single graph, clearly indicating the value of V_{GS} next to each curve.

PART 3: POST-MEASUREMENT EXERCISE

Simulation vs. measurement

What are the main differences between your simulated and measured curves? Can you explain the differences?

Parameter extraction

(1) Threshold voltage, V_{tn}

From the measured I_D vs. V_{GS} curve, at what value of V_{GS} does the NMOS turn on? Set this as the threshold voltage V_{tn}, of your transistor.

(2) MOSFET transconductance parameter, k_n

Based on the value of drain current I_D at V_{GS} = 3.0 V, and using the saturation model for the transistor, i.e., $I_D = (1/2)k_n(V_{GS} - V_{tn})^2$, extract the value of $k_n = \mu_n C_{ox}(W/L)$. Using your extracted values of V_{tn} and k_n, plot a curve of I_D vs. V_{GS}, using the saturation model, and compare with your simulated and measured curves. Are there any differences? Can you explain the differences?

(3) Early voltage, V_A

Based on your measured I_D vs. V_{DS} curves for a saturated transistor, extract the Early voltage, V_A. Does V_A change significartly for each value of V_{GS}? What is the average value of V_A? Based on your average value of V_A, calculate $\lambda_n = 1/V_A$.

Repeat Steps 1 to 3 for your measured results.

Summarize your results in the following table.

	MEASURED
V_{tn} [V]	
k_n [mA/V²]	
λ_n [V⁻¹]	

PART 4 [OPTIONAL]: EXTRA EXPLORATION

If you have access to a semiconductor parameter analyzer, generate the I_D vs. V_{DS} curves using the analyzer. How do they compare to the curves you generated in Part 3? Re-extract values of V_{tn}, k_n, and λ_n.

PMOS I-V Characteristics
(See Sections 5.1–5.2, p. 246 of Sedra/Smith)

OBJECTIVES:

To study PMOS transistor I-V curves by:

- Simulating a transistor to investigate the drain current vs. gate-to-source voltage and drain-to-source voltage.
- Implementing a circuit and taking measurements of the I_D vs. V_{SG} and I_D vs. V_{SD} curves.
- Extracting values of k_p, V_{tp}, and λ_p.

MATERIALS:

- Laboratory setup, including breadboard
- 1 enhancement-type PMOS transistor (e.g., MC14007)
- Several wires

PART 1: SIMULATION

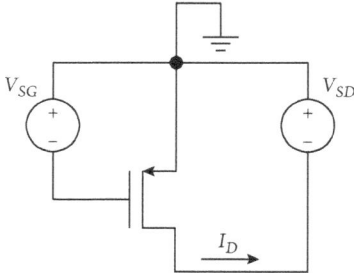

FIGURE L5.2: Transistor measurement circuit. See Table 5.1 in S&S.

Consider the circuit in Figure L5.2. Enter the circuit into your simulator's schematic editor, applying DC voltage supplies to the gate and drain of the transistor. In the diagram, the source is indicated as the reference node (ground). What voltages would you need to apply if another node, e.g., the drain, were labeled as the reference?

I_D vs. V_{SG}

While setting V_{SD} to a constant value of 5 V, sweep the gate voltage from 0 V to –5 V in increments of 0.1 V. Plot a curve of I_D vs. V_{SG}. At what value of V_{SG} does the PMOS turn on?

I_D vs. V_{SD}

For three values of V_{SG} (2.5 V, 3.0 V, and 3.5 V), sweep the drain voltage from 0 V to –5 V in increments of 0.1 V. Plot the curves for I_D vs. V_{SD} onto a single graph, clearly indicating the value of V_{SG} next to each curve.

PART 2: MEASUREMENTS

Assemble the circuit from Figure L5.2, using a power supply to generate the DC voltages. Note the polarities of the voltage sources. You may need to be creative to get the correct polarities! Remember that for a PMOS transistor that is on, V_{SG}, V_{SD}, and I_D will be positive quantities.

I_D vs. V_{SG}

While setting V_{SD} to a constant value of 5 V, sweep the gate voltage from –1.0 V to –3.5 V in increments of 0.25 V (note, we have reduced the number of data points with respect to the simulations), and measure the drain current using the power supply. (*Note*: Not all power supplies allow you to measure current accurately; if this is the case for your lab setup, you may place a small resistor in series with the drain and measure the voltage drop across the resistor.) Plot a curve of I_D vs. V_{SG}. At what value of V_{SG} does the current turn on?

I_D vs. V_{SD}

For three values of V_{GS} (2.5 V, 3.0 V, and 3.5 V), sweep the drain voltage from 0.0 V to –3.5 V in increments of 0.5 V, and measure the drain current using the power supply. Plot the curves for I_D vs. V_{SD} onto a single graph, clearly indicating the value of V_{SG} next to each curve.

PART 3: POST-MEASUREMENT EXERCISE

Simulation vs. measurement

What are the main differences between your simulated and measured curves? Can you explain the differences?

Parameter extraction

[1] Threshold voltage, V_{tp}

From the measured I_D vs. V_{SG} curve, at what value of V_{SG} does the PMOS turn on? Set this as the threshold voltage V_{tp} of your transistor, but express it as a negative number to be consistent with practice.

(2) MOSFET transconductance parameter, k$_p$

Based on the value of drain current I_D at $V_{SG} = -V_{tp} + 1$ V, and using the saturation model for the transistor, i.e., $I_D = (1/2)k_p(V_{SG} - |V_{tp}|)^2$, extract the value of $k_p = \mu_p C_{ox}(W/L)$. Using your extracted values of V_{tp} and k_p, plot a curve of I_D vs. V_{SG}, using the saturation model, and compare with your simulated and measured curves. Are there any differences? Can you explain the differences?

(3) Early voltage, V$_A$

Based on your measured I_D vs. V_{SD} curves for a saturated transistor, extract the Early voltage V_A. Does V_A change significantly for each value of V_{SG}? What is the average value of V_A? Based on your average value of V_A, calculate $\lambda_p = 1/V_A$.

Repeat Steps 1 to 3 for your measured results.

Summarize your results in the following table.

	MEASURED
V_{tp} [V]	
k_p [mA/V^2]	
λ_p [V^{-1}]	

PART 4 [OPTIONAL]: EXTRA EXPLORATION

If you have access to a semiconductor parameter analyzer, generate the I_D vs. V_{SD} curves using the analyzer. How do they compare to the curves you generated in Part 3? Re-extract values of V_{tp}, k_p, and λ_p.

NMOS at DC
(See Section 5.3, p. 273 of Sedra/Smith)

OBJECTIVES:

To study DC biasing of an NMOS transistor by:

- Completing the DC analysis of three circuits: (1) an NMOS biased in the saturation region, (2) an NMOS biased in the triode region, and (3) a diode-connected NMOS transistor.
- Simulating the circuits to compare the results with the paper analysis.
- Implementing the circuits in an experimental setting, taking measurements, and comparing their performance to theoretical and simulated results.
- Qualitatively seeing the impact of transistor-to-transistor variations.

MATERIALS:

- Laboratory setup, including breadboard
- 1 enhancement-type NMOS transistor (e.g., MC14007)
- Several wires and resistors of varying sizes

PART 1: NMOS IN SATURATION MODE

Consider the circuit shown in Figure L5.3.1:

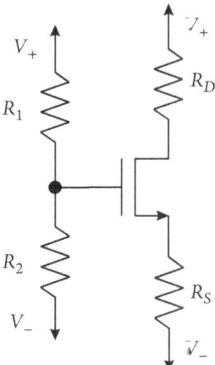

FIGURE L5.3.1: NMOS-based circuit for saturation- and triode-mode operation. Related to Fig. 5.24 in S&S.

Design the circuit in Figure L5.3.1 such that $I_D = 1$ mA, $V_G = 0$ V, and $V_D = +5$ V. Use supplies of $V_+ = -V_- = 15$ V.

Hand calculations

- Sketch the circuit in your lab book, clearly labeling the transistor's three terminals.
- Based on the specifications, calculate V_{OV}.
- From the datasheet, find the threshold voltage V_{tn} of the transistor (*Note*: The datasheet may use a different symbol, e.g., V_{TH} ...) or alternately use your value from LAB 5.1. What is V_{GS}? (*Note*: The datasheet indicates a range of values of V_{tn} that you will find among the batch of transistors. Use the "nominal" value in your calculations, but remember: Your actual transistor has a value of V_{tn} that falls somewhere in that range, which will slightly affect your measurement results!) What is V_S?
- You now have enough information to calculate R_S. Is the calculated value available in your kit? Can you achieve this value by combining several resistors? Comment.
- You also have enough information to calculate R_D. Is the calculated value available in your kit? Can you achieve this value by combining several resistors? Comment.
- What values of R_1 and R_2 do you need to use? Is the problem completely specified?

Simulation

- Simulate your circuit using the values of R_S, R_D, R_1, and R_2 based on your preceding calculations.
- Report the values of V_S, V_D, V_G, and I_D. How closely do they match your calculations? (Remember: The simulator has its own, more complex model of the real transistor, so there should be some small variations.)

Prototyping and Measurement

- Assemble the circuit onto a breadboard.
- Using a digital multimeter, measure V_G, V_S, and V_D.
- Using a digital multimeter, measure all resistors to three significant digits.

Post-Measurement Exercise

- What are the measured values of V_{GS} and V_{DS}? How do they compare to your pre-lab calculations? Explain any discrepancies.
- Based on the measured values of V_D and V_S and your measured resistor values, what is the measured value of I_D based on your lab measurements?

PART 2: NMOS IN TRIODE MODE

Redesign the circuit in Figure L5.3.1 such that $I_D = 1$ mA, $V_D = +2$ V, and $V_{DS} = 0.5$ V. Use supplies of $V_+ = -V_- = 15$ V. Note that you must use the triode model.

Hand calculations

- Sketch the circuit in your lab book, clearly labeling the transistor's three terminals.
- Based on the specifications, calculate V_{OV} and V_{GS}. What is V_G?
- You now have enough information to calculate R_S. Is the calculated value available in your kit? Can you achieve this value by combining several resistors? Comment.
- You also have enough information to calculate R_D. Is the calculated value available in your kit? Can you achieve this value by combining several resistors? Comment.
- What values of R_1 and R_2 do you need to use? Is the problem completely specified?

Simulation

- Simulate your circuit using the values of R_S, R_D, R_1, and R_2 based on your calculations.
- Report the values of V_S, V_D, V_G, and I_D. How closely do they match your calculations?

Prototyping and Measurement

- Assemble the circuit onto a breadboard.
- Using a digital multimeter, measure V_G, V_S, and V_D. Report them in your lab book.
- Using a digital multimeter, measure all resistors to three significant digits.

Post-Measurement Exercise

- What are the measured values of V_{GS} and V_{DS}? How do they compare to your pre-lab calculations? Explain any discrepancies. Is the transistor in the triode operating region?
- Based on the measured values of V_D and V_S and your measured resistor values, what is the measured value of I_D based on your lab measurements?

PART 3: DIODE-CONNECTED NMOS

Consider the circuit shown in Figure L5.3.2:

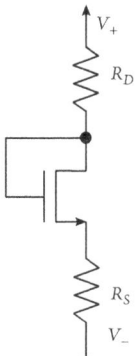

FIGURE L5.3.2: Circuit with diode-connected NMOS transistor.

Design the circuit in Figure L5.3.2 such that $I_D = 1$ mA and $R_S = 15$ kΩ. Use supplies of $V_+ = -V_- = 15$ V.

Hand calculations
- Sketch the circuit in your lab book, clearly labeling the transistor's three terminals.
- What is the operating region of the transistor? Based on the specifications, calculate V_{OV}. What are V_S and V_D?
- You now have enough information to calculate R_D. Is the calculated value available in your kit? Can you achieve this value by combining several resistors? Comment.

Simulation
- Simulate your circuit using the value R_D based on your calculations.
- Report the values of V_S, V_D, and I_D. How closely do they match your calculations?

Prototyping and Measurement
- Assemble the circuit onto a breadboard.
- Using a digital multimeter, measure V_S and V_D. Report them in your lab book.
- Using a digital multimeter, measure all resistors to three significant digits.

Post-Measurement Exercise
- How do the measured values compare to your pre-lab calculations? Explain any discrepancies.
- Based on the measured values of V_D and V_S and your measured resistor values, what is the measured value of I_D based on your lab measurements?

PART 4 [OPTIONAL]: EXTRA EXPLORATION

- In this exploration, reuse the circuit from Part 1 as well as the values of R_S and R_D that you calculate for that circuit. However, replace the R_1-R_2 voltage divider with a 10-kΩ 20-turn potentiometer, with the central pin connected to the transistor gate. This will allow you to adjust the DC voltage at the gate.
- Gradually increase the gate voltage, and make recordings of V_G, V_D, and V_S as you sweep the gate voltage. What do you observe? Are the trends what you expect them to be? Can you indicate the transition points between the different transistor operating regions?

LAB 5.4

PMOS at DC
(See Section 5.3, p. 273 of Sedra/Smith)

OBJECTIVES:

To study DC biasing of a PMOS transistor by:

- Completing the DC analysis of three circuits: (1) a PMOS biased in the saturation region, (2) a PMOS biased in the triode region, and (3) a diode-connected PMOS transistor.
- Simulating the circuits to compare the results with the paper analysis.
- Implementing the circuits in an experimental setting, taking measurements, and comparing their performance to theoretical and simulated results.
- Qualitatively seeing the impact of transistor-to-transistor variations.

MATERIALS:

- Laboratory setup, including breadboard
- 1 enhancement-type PMOS transistor (e.g., MC14007)
- Several wires and resistors of varying sizes

PART 1: PMOS IN SATURATION MODE

Consider the circuit shown in Figure L5.4.1:

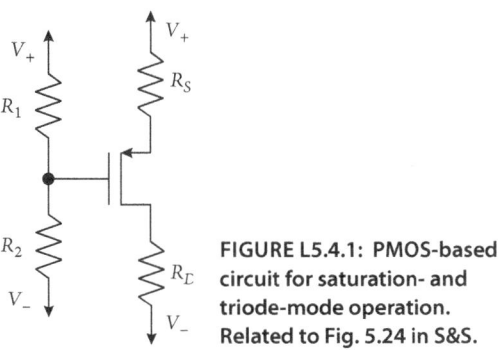

FIGURE L5.4.1: PMOS-based circuit for saturation- and triode-mode operation. Related to Fig. 5.24 in S&S.

Design the circuit in Figure L5.4.1 such that $I_D = 1$ mA, $V_G = 0$ V, and $V_D = -5$ V. Use supplies of $V_+ = -V_- = 15$ V.

Hand calculations

- Sketch the circuit in your lab book, clearly labeling the transistor's three terminals.
- Based on the specifications, calculate V_{OV}.
- From the datasheet, find the threshold voltage V_{tp} of the transistor (*Note*: The datasheet may use a different symbol, e.g., V_{TH} . . .), or alternately use your value from LAB 5.2. What is V_{SG}? (*Note*: The datasheet indicates a *range* of values of V_{tp} that you will find among the batch of transistors. Use the "nominal" value in your calculations, but remember: Your actual transistor has a value of V_{tp} that falls somewhere in that range, which will affect your measurement results slightly!) What is V_S?
- You now have enough information to calculate R_S. Is the calculated value available in your kit? Can you achieve this value by combining several resistors? Comment.
- You also have enough information to calculate R_D. Is the calculated value available in your kit? Can you achieve this value by combining several resistors? Comment.
- What values of R_1 and R_2 do you need to use? Is the problem completely specified?

Simulation

- Simulate your circuit using the values of R_S, R_D, R_1, and R_2 based on your preceding calculations.
- Report the values of V_S, V_D, V_G, and I_D. How closely do they match your calculations? (Remember: The simulator has its own, more complex model of the real transistor, so there should be some small variations.)

Prototyping and Measurement

- Assemble the circuit onto a breadboard.
- Using a digital multimeter, measure V_G, V_S, and V_D.
- Using a digital multimeter, measure all resistors to three significant digits.

Post-Measurement Exercise

- What are the measured values of V_{SG} and V_{SD}? How do they compare to your pre-lab calculations? Explain any discrepancies.
- Based on the measured values of V_D and V_S and your measured resistor values, what is the measured value of I_D based on your lab measurements?

PART 2: PMOS IN TRIODE MODE

Redesign the circuit in Figure L5.4.1 such that $I_D = 1$ mA, $V_D = -2$ V, and $V_{SD} = 0.5$ V. Use supplies of $V_+ = -V_- = 15$ V. Note that you must use the triode model.

Hand calculations

- Sketch the circuit in your lab book, clearly labeling the transistor's three terminals.
- Based on the specifications, calculate V_{OV} and V_{SG}. What is V_G?

- You now have enough information to calculate R_S. Is the calculated value available in your kit? Can you achieve this value by combining several resistors? Comment.
- You also have enough information to calculate R_D. Is the calculated value available in your kit? Can you achieve this value by combining several resistors? Comment.
- What values of R_1 and R_2 do you need to use? Is the problem completely specified?

Simulation

- Simulate your circuit using the values of R_S, R_D, R_1, and R_2 based on your calculations.
- Report the values of V_S, V_D, V_G, and I_D. How closely do they match your calculations?

Prototyping and Measurement

- Assemble the circuit onto a breadboard.
- Using a digital multimeter, measure V_G, V_S, and V_D. Report them in your lab book.
- Using a digital multimeter, measure all resistors to three significant digits.

Post-Measurement Exercise

- What are the measured values of V_{SG} and V_{SD}? How do they compare to your pre-lab calculations? Explain any discrepancies. Is the transistor in the triode operating region?
- Based on the measured values of V_D and V_S and your measured resistor values, what is the measured value of I_D based on your lab measurements?

PART 3: DIODE-CONNECTED PMOS

Consider the circuit shown in Figure L5.4.2:

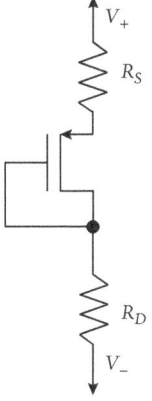

FIGURE L5.4.2: Circuit with diode-connected PMOS transistor.

Design the circuit in Figure L5.4.2 such that $I_D = 1$ mA and $R_S = 15$ kΩ. Use supplies of $V_+ = -V_- = 15$ V.

Hand calculations

- Sketch the circuit in your lab book, clearly labeling the transistor's three terminals.
- What is the operating region of the transistor? Based on the specifications, calculate V_{OV}. What are V_S and V_D?
- You now have enough information to calculate R_D. Is the calculated value available in your kit? Can you achieve this value by combining several resistors? Comment.

Simulation

- Simulate your circuit using the value R_D based on your calculations.
- Report the values of V_S, V_D, and I_D. How closely do they match your calculations?

Prototyping and Measurement

- Assemble the circuit onto a breadboard.
- Using a digital multimeter, measure V_S and V_D. Report them in your lab book.
- Using a digital multimeter, measure all resistors to three significant digits.

Post-Measurement Exercise

- How do the measured values compare to your pre-lab calculations? Explain any discrepancies.
- Based on the measured values of V_D and V_S and your measured resistor values, what is the measured value of I_D based on your lab measurements?

PART 4 [OPTIONAL]: EXTRA EXPLORATION

- In this exploration, reuse the circuit from Part 1 as well as the values of R_S and R_D that you calculate for that circuit. However, replace the R_1-R_2 voltage divider with a 10-kΩ 20-turn potentiometer, with the central pin connected to the transistor gate. This will allow you to adjust the DC voltage at the gate.
- Gradually increase the gate voltage, and make recordings of V_G, V_D, and V_S as you sweep the gate voltage. What do you observe? Are the trends what you expect them to be? Can you indicate the transition points between the different transistor operating regions?

NPN I-V Characteristics
(See Sections 6.1–6.2, p. 305 of Sedra/Smith)

OBJECTIVES:

To study NPN transistor I-V curves by:

- Simulating a transistor to investigate the collector current vs. base-to-emitter voltage and collector-to-emitter voltage.
- Implementing a circuit and taking measurements of the I_C vs. V_{BE} and I_C vs. V_{CE} curves.
- Extracting values of β and V_A.

MATERIALS:

- Laboratory setup, including breadboard
- 1 NPN transistor (e.g., NTE2321)
- Several wires

PART 1: SIMULATION

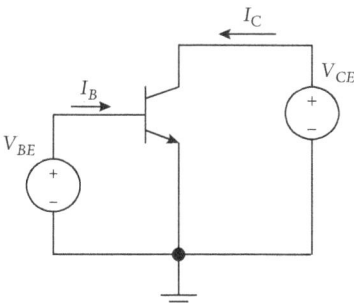

FIGURE L6.1: Transistor measurement circuit. Related to Fig. 6.13 in S&S.

Consider the circuit in Figure L6.1. Enter the circuit into your simulator's schematic editor, applying DC voltage supplies to the base and collector of the transistor.

I_C vs. V_{BE}

While setting V_{CE} to a constant value of 5 V, sweep the base voltage from 0 V to 0.8 V in increments of 0.1 V. Plot a curve of I_C vs. V_{BE}. At what value of V_{BE} does current begin to conduct? What are the values of I_B and I_C when $V_{BE} = 0.7$ V? Based on these numbers, what is your estimate of β?

I_C vs. V_{CE}

For three values of V_{BE} (0.6 V, 0.7 V, and 0.8 V), sweep the collector voltage from 0 V to 2 V in increments of 0.1 V. Plot the curves for I_C vs. V_{CE} onto a single graph, clearly indicating the value of V_{GS} next to each curve.

PART 2: MEASUREMENTS

Assemble the circuit from Figure L6.1, using a power supply to generate the DC voltages.

I_C vs. V_{BE}

While setting V_{CE} to a constant value of 5 V, sweep the base voltage from 0 V to 0.8 V in increments of 0.1 V, and measure the collector current using the power supply. (Note: Not all power supplies allow you to accurately measure current; if this is the case for your lab setup, you may place a small resistor in series with the collector and measure the voltage drop across the resistor.) Plot a curve of I_C vs. V_{BE}. At what value of V_{BE} does the current turn on? Using small resistors placed in series with the base and collector terminals, measure I_B and I_C for $V_{BE} = 0.7$ V? Based on these numbers, what is your estimate of β?

I_C vs. V_{CE}

For three values of V_{BE} (0.6 V, 0.7 V, and 0.8 V), sweep the V_{CE} from 0 V to 1 V in increments of 0.1 V, and measure the collector current using the power supply. Plot the curves for I_C vs. V_{CE} onto a single graph, clearly indicating the value of V_{BE} next to each curve.

PART 3: POST-MEASUREMENT EXERCISE

Simulation vs. measurement

What are the main differences between your simulated and measured curves? Can you explain the differences?

Early voltage, V_A

Based on your simulated I_C vs. V_{CE} curves for an active transistor, extract the Early voltage, V_A. Does V_A change significantly for each value of V_{BE}? What is the average value of V_A?

PART 4 [OPTIONAL]: EXTRA EXPLORATION

If you have access to a semiconductor parameter analyzer, generate the I_C vs. V_{CE} curves using the analyzer. How do they compare to the curves you generated in Part 3? Re-extract values of β and V_A.

PNP I-V Characteristics
(See Sections 6.1–6.2, p. 305 of Sedra/Smith)

OBJECTIVES:

To study PNP transistor I-V curves by:

*   Simulating a transistor to investigate the collector current vs. base-to-emitter voltage and collector-to-emitter voltage.
*   Implementing a circuit and taking measurements of the I_C vs. V_{EB} and I_C vs. V_{EC} curves.
*   Extracting values of β and V_A.

MATERIALS:

*   Laboratory setup, including breadboard
*   1 PNP transistor (e.g., NTE2322)
*   Several wires

PART 1: SIMULATION

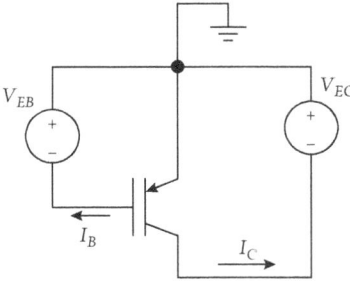

FIGURE L6.2: Transistor measurement circuit. Related to Fig. 6.13 in S&S.

Consider the circuit in Figure L6.2. Enter the circuit into your simulator's schematic editor, applying DC voltage supplies to the base and collector of the transistor. In the diagram, the emitter is indicated as the reference node (ground). What voltages would you need to apply if another node, e.g., the collector, were labeled as the reference?

I_C vs. V_{EB}

> While setting V_{EC} to a constant value of 5 V, sweep the base voltage from 0 V to –0.8 V in increments of 0.1 V. Plot a curve of I_C vs. V_{EB}. At what value of V_{EB} does current begin to conduct? What are the values of I_B and I_C when $V_{EB} = 0.7$ V? Based on these numbers, what is your estimate of β?

I_C vs. V_{EC}

> For three values of V_{EB} (0.6 V, 0.7 V, and 0.8 V), sweep the collector voltage from 0 V to –2 V in increments of 0.1 V. Plot the curves for I_C vs. V_{EC} onto a single graph, clearly indicating the value of V_{EB} next to each curve.

PART 2: MEASUREMENTS

> Assemble the circuit from Figure L6.2, using a power supply to generate the DC voltages. You may need to be creative to get the correct polarities! Remember that for a PNP transistor that is on, V_{EB}, V_{EC}, and I_C will be positive quantities, so the base and collector voltages will be negative.

I_C vs. V_{EB}

> While setting V_{EC} to a constant value of 5 V, sweep the base voltage from 0 V to –0.8 V in increments of 0.1 V, and measure the collector current using the power supply. Plot a curve of I_C vs. V_{EB}. (*Note*: Not all power supplies allow you to measure current accurately; if this is the case for your lab setup, you may place a small resistor in series with the collector and measure the voltage drop across the resistor.) At what value of V_{EB} does the current turn on? Using small resistors placed in series with the base and collector terminals, measure I_B and I_C for $V_{EB} = 0.7$ V. Based on these numbers, what is your estimate of β?

I_C vs. V_{EC}

> For three values of V_{EB} (0.6 V, 0.7 V, and 0.8 V), sweep V_{EC} from 0 V to 1 V in increments of 0.1 V, and measure the collector current using the power supply. Plot the curves for I_C vs. V_{EC} onto a single graph, clearly indicating the value of V_{EB} next to each curve.

PART 3: POST-MEASUREMENT EXERCISE

Simulation vs. measurement

> What are the main differences between your simulated and measured curves? Can you explain the differences?

Early voltage, V_A

> Based on your simulated I_C vs. V_{EC} curves for an active transistor, extract the Early voltage, V_A. Does V_A change significantly for each value of V_{EB}? What is the average value of V_A?

PART 4 [OPTIONAL]: EXTRA EXPLORATION

If you have access to a semiconductor parameter analyzer, generate the I_C vs. V_{EC} curves using the analyzer. How do they compare to the curves you generated in Part 3? Re-extract values of β and V_A.

NPN at DC
(See Section 6.3, p. 333 of Sedra/Smith)

OBJECTIVES:

To study DC biasing of an NPN bipolar transistor by:

- Completing the DC analysis of three circuits: (1) an NPN transistor that is biased in the active region, (2) an NPN transistor that is biased in the saturation region, and (3) a diode-connected NPN transistor.
- Simulating the circuits to compare the results with the paper analysis.
- Implementing the circuits in an experimental setting, taking measurements, and comparing their performance with theoretical and simulated results.
- Qualitatively seeing the impact of transistor-to-transistor variations.

MATERIALS:

- Laboratory setup, including breadboard
- 1 NPN transistor (e.g., NTE2321)
- Several wires and resistors of varying sizes

PART 1: NPN IN ACTIVE MODE

Consider the circuit shown in Figure L6.3.1:

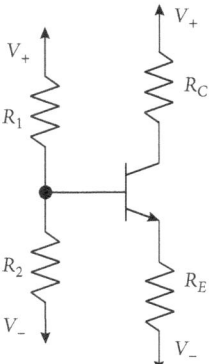

FIGURE L6.3.1: NPN-based circuit. Related to Fig. 6.29(a) in S&S.

Design the circuit in Figure L6.3.1 such that $I_C = 1$ mA, $V_B = 0$ V, and $V_C = +5$ V. Use supplies of $V_+ = -V_- = 15$ V. Use $\beta = 100$.

Hand calculations

- Sketch the circuit in your lab book, clearly labeling the transistor's three terminals.
- What are I_B and I_E? Based on these numbers, what is V_E?
- You now have enough information to calculate R_E and R_C. Are the calculated values available in your kit? Can you achieve these values by combining several resistors? Comment.
- Derive the Thévenin equivalent of R_1 and R_2. What values of R_1 and R_2 do you need to use to achieve $V_B = 0$ V? Remember that $I_B \neq 0$. Is the problem completely specified? If not, what needs to be specified?

Simulation

- Simulate your circuit using values of R_E, R_C, R_1, and R_2 based on your calculations.
- Report the values of V_E, V_C, V_B, I_E, I_C, and I_B. How closely do they match your calculations? (Remember: The simulator has its own, more complex model of the real transistor, so there should be some small variations.)

Prototyping and Measurement

- Assemble the circuit onto a breadboard.
- Using a digital multimeter, measure V_E, V_C, and V_B.
- Using a digital multimeter, measure all resistors to three significant digits.

Post-Measurement Exercise

- What are the measured values of V_{BE} and V_{CE}? How do they compare to your pre-lab calculations? Explain any discrepancies.
- Based on the measured values of V_C and V_E and your measured resistor values, what are the measured values of I_B, I_C, and I_E based on your lab measurements?

PART 2: NPN IN SATURATION MODE

Redesign the circuit in Figure L6.3.1 such that $I_C = 1$ mA, $I_E = 1.2$ mA, $V_C = +2$ V, and $V_{CE} = 0.2$ V. Use supplies of $V_+ = -V_- = 15$ V. Note that you must use the saturation model.

Hand calculations

- Sketch the circuit in your lab book, clearly labeling the transistor's three terminals.
- Based on the specifications, calculate V_E and V_B.
- You now have enough information to calculate R_C and R_E. Are the calculated values available in your kit? Can you achieve this value by combining several resistors? Comment.
- What is β_{forced}?
- What values of R_1 and R_2 do you need to use? Is the problem completely specified?

Simulation

- Simulate your circuit using values of R_C, R_E, R_1, and R_2 based on your calculations.
- Report the values of V_E, V_C, V_B, I_E, I_C, and I_B. How closely do they match your calculations?

Prototyping and Measurement

- Assemble the circuit onto a breadboard.
- Using a digital multimeter, measure V_E, V_C, and V_B. Report them in your lab book.
- Using a digital multimeter, measure all resistors to three significant digits.

Post-Measurement Exercise

- What are the measured values of V_{BE} and V_{CE}? How do they compare to your pre-lab calculations? Explain any discrepancies.
- Based on the measured voltages and resistor values, what are the measured values of I_E, I_C, and I_B based on your lab measurements? What is β_{forced}?

PART 3: DIODE-CONNECTED NPN

Consider the circuit shown in Figure L6.3.2:

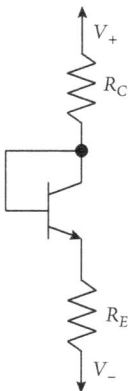

FIGURE L6.3.2: Circuit with diode-connected NPN transistor.

Design the circuit in Figure L6.3.2 such that $I_C = 1$ mA and $R_E = 15$ kΩ. Use supplies of $V_+ = -V_- = 15$ V. Use $\beta = 100$.

Hand calculations

- Sketch the circuit in your lab book, clearly labeling the transistor's three terminals.
- What is the operating region of the transistor? Calculate V_C.

- You now have enough information to calculate R_C. Is the calculated value available in your kit? Can you achieve this value by combining several resistors? Comment.

Simulation

- Simulate your circuit using the value R_C based on your calculations.
- Report the values of V_C, V_E, I_B, I_C, and I_E. How closely do they match your calculations?

Prototyping and Measurement

- Assemble the circuit onto a breadboard.
- Using a digital multimeter, measure V_C and V_E. Report them in your lab book.
- Using a digital multimeter, measure all resistors to three significant digits.

Post-Measurement Exercise

- How do the measured values compare to your pre-lab calculations? Explain any discrepancies.
- Based on the measured values of V_C and V_E and your measured resistor values, what are the measured values of I_E, I_C, and I_B based on your lab measurements?

PART 4 [OPTIONAL]: EXTRA EXPLORATION

- In this exploration, reuse the circuit from Part 1 as well as the values of R_E and R_C that you calculate for that circuit. However, replace the R_1-R_2 voltage divider with a 10-kΩ 20-turn potentiometer, with the central pin connected to the transistor base. This will allow you to adjust the DC voltage at the base.
- Gradually increase the base voltage, and make recordings of V_B, V_C, and V_E as you sweep the base voltage. What do you observe? Are the trends what you expect them to be? Can you indicate the transition points between the different transistor operating regions?

PNP at DC
(See Section 6.3, p. 333 of Sedra/Smith)

OBJECTIVES:

To study DC biasing of a PNP bipolar transistor by:

- Completing the DC analysis of three circuits: (1) a PNP transistor that is biased in the active region, (2) a PNP transistor that is biased in the saturation region, and (3) a diode-connected PNP transistor.
- Simulating the circuits to compare the results with the paper analysis.
- Implementing the circuits in an experimental setting, taking measurements, and comparing their performance with theoretical and simulated results.
- Qualitatively seeing the impact of transistor-to-transistor variations.

MATERIALS:

- Laboratory setup, including breadboard
- 1 PNP transistor (e.g., NTE2322)
- Several wires and resistors of varying sizes

PART 1: PNP IN ACTIVE MODE

Consider the circuit shown in Figure L6.4.1:

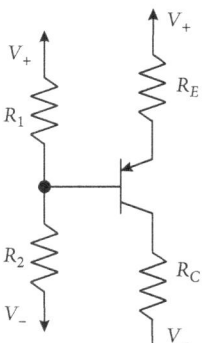

FIGURE L6.4.1: PNP-based circuit. Related to Fig. 6.29(a) in S&S.

Design the circuit in Figure L6.4.1 such that $I_C = 1$ mA, $V_B = 0$ V, and $V_C = -5$ V. Use supplies of $V_+ = -V_- = 15$ V. Use $\beta = 100$.

44

Hand calculations

- Sketch the circuit in your lab book, clearly labeling the transistor's three terminals.
- What are I_B and I_E? Based on these numbers, what is V_E?
- You now have enough information to calculate R_E and R_C. Are the calculated values available in your kit? Can you achieve these values by combining several resistors? Comment.
- Derive the Thévenin equivalent of R_1 and R_2. What values of R_1 and R_2 do you need to use to achieve $V_B = 0$ V? Remember that $I_B \neq 0$. Is the problem completely specified?

Simulation

- Simulate your circuit using values of R_E, R_C, R_1, and R_2 based on your calculations.
- Report the values of V_E, V_C, V_B, I_E, I_C, and I_B. How closely do they match your calculations? (Remember: The simulator has its own, more complex model of the real transistor, so there should be some small variations.)

Prototyping and Measurement

- Assemble the circuit onto a breadboard.
- Using a digital multimeter, measure V_E, V_C, and V_B.
- Using a digital multimeter, measure all resistors to three significant digits.

Post-Measurement Exercise

- What are the measured values of V_{EB} and V_{EC}? How do they compare to your pre-lab calculations? Explain any discrepancies.
- Based on the measured values of V_C and V_E and your measured resistor values, what are the measured values of I_B, I_C, and I_E based on your lab measurements?

PART 2: PNP IN SATURATION MODE

Redesign the circuit in Figure L6.4.1 such that $I_C = 1$ mA, $I_E = 1.2$ mA, $V_C = -2$ V, and $V_{EC} = 0.2$ V. Use supplies of $V_+ = -V_- = 15$ V. Note that you must use the saturation model.

Hand calculations

- Sketch the circuit in your lab book, clearly labeling the transistor's three terminals.
- Based on the specifications, calculate V_E and V_B.
- You now have enough information to calculate R_C and R_E. Are the calculated values available in your kit? Can you achieve this value by combining several resistors? Comment.
- What is β_{forced}?
- What values of R_1 and R_2 do you need to use? Is the problem completely specified?

Simulation

- Simulate your circuit using values of R_C, R_E, R_1, and R_2 based on your calculations.
- Report the values of V_E, V_C, V_B, I_E, I_C, and I_B. How closely do they match your calculations?

Prototyping and Measurement

- Assemble the circuit onto a breadboard.
- Using a digital multimeter, measure V_E, V_C, and V_B. Report them in your lab book.
- Using a digital multimeter, measure all resistors to three significant digits.

Post-Measurement Exercise

- What are the measured values of V_{EB} and V_{EC}? How do they compare to your pre-lab calculations? Explain any discrepancies.
- Based on the measured voltages and resistor values, what are the measured values of I_E, I_C, and I_B based on your lab measurements? What is β_{forced}?

PART 3: DIODE-CONNECTED PNP

Consider the circuit shown in Figure L6.4.2:

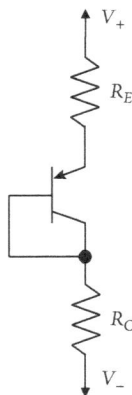

FIGURE L6.4.2: Circuit with diode-connected PNP transistor.

Design the circuit in Figure L6.4.2 such that $I_C = 1$ mA and $R_E = 15$ kΩ. Use supplies of $V_+ = -V_- = 15$ V. Use $\beta = 100$.

Hand calculations

- Sketch the circuit in your lab book, clearly labeling the transistor's three terminals.
- What is the operating region of the transistor? Calculate V_C.

- You now have enough information to calculate R_C. Is the calculated value available in your kit? Can you achieve this value by combining several resistors? Comment.

Simulation

- Simulate your circuit using the value of R_C based on your calculations.
- Report the values of V_C, V_E, I_B, I_C, and I_E. How closely do they match your calculations?

Prototyping and Measurement

- Assemble the circuit onto a breadboard.
- Using a digital multimeter, measure V_C and V_E. Report them in your lab book.
- Using a digital multimeter, measure all resistors to three significant digits.

Post-Measurement Exercise

- How do the measured values compare to your pre-lab calculations? Explain any discrepancies.
- Based on the measured values of V_C and V_E and your measured resistor values, what are the measured values of I_E, I_C, and I_B based on your lab measurements?

PART 4 [OPTIONAL]: EXTRA EXPLORATION

- In this exploration, reuse the circuit from Part 1 as well as the values of R_E and R_C that you calculate for that circuit. However, replace the R_1-R_2 voltage divider with a 10-kΩ 20-turn potentiometer, with the central pin connected to the transistor base. This will allow you to adjust the DC voltage at the base.
- Gradually increase the base voltage, and make recordings of V_B, V_C, and V_E as you sweep the base voltage. What do you observe? Are the trends what you expect them to be? Can you indicate the transition points between the different transistor operating regions?

NMOS Common-Source Amplifier
(See Section 7.5.1, p. 461 of Sedra/Smith)

OBJECTIVES:

To study an NMOS-based common-source (CS) amplifier by:

- Completing the DC and small-signal analysis based on its theoretical behavior.
- Simulating it to compare the results with the paper analysis.
- Implementing it in an experimental setting, taking measurements, and comparing its performance with theoretical and simulated results.
- Measuring its output resistance.
- Qualitatively seeing the impact of transistor-to-transistor variations.

MATERIALS:

- Laboratory setup, including breadboard
- 1 enhancement-type NMOS transistor (e.g., MC14007)
- 3 large (e.g., 47-μF) capacitors
- Several resistors of varying sizes
- Wires

PART 1: DESIGN AND SIMULATION

Consider the circuit shown in Figure L7.1:

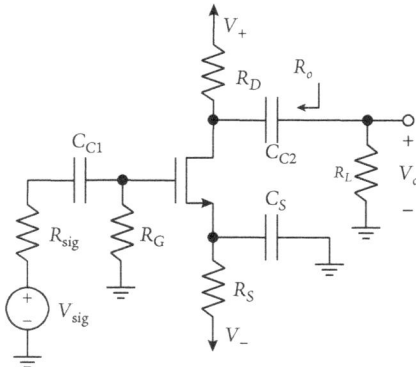

FIGURE L7.1: Common-source amplifier circuit, with coupling capacitors, and resistor R_G for DC-biasing purposes. Based on Fig. 7.57 p. 462 S&S.

Design the amplifier to achieve a small-signal gain of at least $A_v = -5$ V/V. Use supplies of $V_+ = -V_- = 15$ V, $R_{sig} = 50$ Ω, $R_L = 10$ kΩ, $R_G = 10$ kΩ, and design the circuit to have $I_D = 1$ mA. Obtain the datasheet for the NMOS transistor that will be used. In your lab book, perform the following:

DC Operating Point Analysis

- Sketch a DC model of the circuit in your lab book, replacing the three "large-valued" coupling capacitors—C_{C1}, C_{C2}, C_S—by open circuits (for simplicity you may also omit v_{sig}, R_{sig}, and R_L). What is the DC current through R_G?
- Based on the information just given, you have enough information to calculate $V_{OV} = V_{GS} - V_{tn}$. What is its value? What is the value of g_m? What is V_{GS}? Remember: Your actual transistor will have a value of V_{tn} that will vary from its nominal value, which will alter your measurement results slightly!
- Calculate r_o.
- You now have enough information to calculate R_S. Show your calculations. Is the value you calculate for R_S available in your kit? Can you achieve this value by combining several resistors? Comment.
- *Note*: At this stage we know neither V_{DS} nor R_D.

AC Analysis

- Sketch a small-signal model of the circuit in your lab book, replacing the transistor with its small-signal model, replacing the capacitors with short circuits (what happens to R_S?), and replacing V_+ with an AC ground. What happens to V_-? Label the gate of the transistor as v_i, i.e., the small-signal voltage at the input.
- What is the ratio of v_i/v_{sig}? How would you approximate it in further calculations?
- Derive an expression for $A_v = v_o/v_i$. What is the value of R_D that produces a small-signal voltage gain of *at least* $A_v = -5$ V/V? Is the value you calculated for R_D available in your kit? Can you achieve this value by combining several resistors? Comment.
- What is the DC voltage at the drain? Does this satisfy the assumption that the transistor should be operating in the saturation region? Explain.
- What is the output resistance, R_o?

Simulation

- Simulate your circuit. Use capacitor values $C_{C1} = C_{C2} = C_S = 47$ μF, and the values of R_S and R_D based on your preceding calculations. Use a 10-mV$_{pk-pk}$, 1-kHz sinusoid with no DC component applied at v_{sig}.
- From your simulation, report the DC values of V_{GS}, V_{DS}, and I_D. How closely do they match your calculations? (Remember: The simulator has its own more-complex model of the real transistor, so there should be some small variations.)
- From your simulation, report A_v. How closely does it match your calculations?

PART 2: PROTOTYPING

- Assemble the circuit onto your breadboard using the specified component values and those just calculated. Note that R_{sig} represents the output

resistance of the function generator, and therefore you should *not* include it in your circuit.

PART 3: MEASUREMENTS

- *DC bias point measurement*: Using a digital multimeter, measure the DC voltages of your circuit at the gate (V_G), source (V_S), and drain (V_D) of your transistor.
- *AC measurement*: Using a function generator, apply to your circuit a 10-mV$_{pk-pk}$, 1-kHz sinusoid with no DC component. (*Note*: Some function generators allow only inputs as small as 50 mV$_{pk-pk}$. If this is the case, use that value instead.)
- Using an oscilloscope, generate plots of v_o and v_i vs. t.
- *Output resistance R_o*: Replace R_L with a 1-MΩ resistor and repeat the AC measurement. What is the amplitude of the output waveform? Adjust R_L until you find a value such that the amplitude of the output waveform is approximately 50% of what it was for the 1-MΩ load. This new value of R_L is the output resistance R_o. How does it compare to the value you calculated earlier in Step 2? *Hint*: It cannot be greater than the value of R_D.
- *Further exploration*: What happens to the shape of the output signal as you increase the amplitude of the input signal, e.g., to 1 V$_{pk-pk}$? At what input amplitude do you begin to see significant distortion? Can you explain this?
- Using a digital multimeter, measure all resistors to three significant digits.

PART 4: POST-MEASUREMENT EXERCISE

- Calculate the values of V_{GS} and V_{DS} that you obtained in the lab. How do they compare to your pre-lab calculations? Explain any discrepancies.
- Based on the measured values of V_D and V_S and your measured resistor values, what is the real value of I_D based on your lab measurements?
- What is the measured value of A_v? How does it compare to your pre-lab calculations? Explain any discrepancies.
- *Hint*: The single biggest source of variations from your pre-lab simulation results will be due to variations in the transistor threshold voltage V_{tn}. Remember: Its value will be somewhere within the range indicated on the transistor datasheet.

PART 5 [OPTIONAL]: EXTRA EXPLORATION

- Instead of tying R_G to ground, try tying it to the drain terminal of the transistor. Repeat the DC bias point measurement and the small-signal gain measurement. What has changed? Do R_D and R_S need to be altered to meet design specifications?

PMOS Common-Source Amplifier
(See Section 7.5.1, p. 461 of Sedra/Smith)

OBJECTIVES:

To study a PMOS-based common-source (CS) amplifier by:

- Completing the DC and small-signal analysis based on its theoretical behavior.
- Simulating it to compare the results with the paper analysis.
- Implementing it in an experimental setting, taking measurements, and comparing its performance with theoretical and simulated results.
- Measuring its output resistance.
- Qualitatively seeing the impact of transistor-to-transistor variations.

MATERIALS:

- Laboratory setup, including breadboard
- 1 enhancement-type PMOS transistor (e.g., MC14007)
- 3 large (e.g., 47-μF) capacitors
- Several resistors of varying sizes
- Wires

PART 1: DESIGN AND SIMULATION

Consider the circuit shown in Figure L7.2:

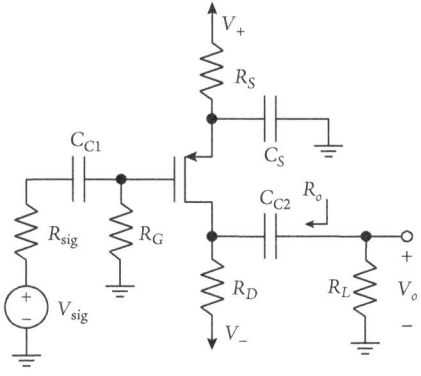

FIGURE L7.2: Common-source amplifier circuit, with coupling capacitors, and resistor R_G for DC-biasing purposes. Based on Fig. 7.57 p. 462 S&S.

Design the amplifier to achieve a small-signal gain of at least $A_v = -5$ V/V. Use supplies of $V_+ = -V_- = 15$ V, $R_{sig} = 50\ \Omega$, $R_L = 10$ kΩ, and $R_G = 10$ kΩ, and design the circuit to have $I_D = 1$ mA. Obtain the datasheet for the PMOS transistor that will be used. In your lab book, perform the following.

DC Operating Point Analysis

- Sketch a DC model of the circuit in your lab book, replacing the three "large-valued" coupling capacitors—C_{C1}, C_{C2}, C_S—by open circuits (for simplicity you may also omit v_{sig}, R_{sig}, and R_L). What is the DC current through R_G?
- Based on the information just given, you have enough information to calculate $V_{OV} = V_{SG} - |V_{tp}|$. What is its value? What is the value of g_m? What is V_{SG}? Remember: Your actual transistor will have a value of V_{tp} that will vary from its nominal value, which will affect your measurement results slightly!
- Calculate r_o.
- You now have enough information to calculate R_S. Show your calculations. Is the value you calculate for R_S available in your kit? Can you achieve this value by combining several resistors? Comment.
- *Note*: At this stage we know neither V_{SD} nor R_D.

AC Analysis

- Sketch a small-signal model of the circuit in your lab book, replacing the transistor with its small-signal model, replacing the capacitors with short circuits (what happens to R_S?), and replacing V_- with an AC ground. What happens to V_+? Label the gate of the transistor as v_i, i.e., the small-signal voltage at the input.
- What is the ratio of v_i/v_{sig}? How would you approximate it in further calculations?
- Derive an expression for $A_v = v_o/v_i$. What is the value of R_D that produces a small-signal voltage gain of *at least* $A_v = -5$ V/V? Is the value you calculated for R_D available in your kit? Can you achieve this value by combining several resistors? Comment.
- What is the DC voltage at the drain? Does this satisfy the assumption that the transistor should be operating in the saturation region? Explain.
- What is the output resistance, R_o?

Simulation

- Simulate the performance of your circuit. Use capacitor values $C_{C1} = C_{C2} = C_S = 47\ \mu$F and the values of R_S and R_D based on your preceding calculations. Use a 10-mV$_{pk-pk}$, 1-kHz sinusoid with no DC component applied at v_{sig}.
- From your simulation, report the DC values of V_{SG}, V_{SD}, and I_D. How closely do they match your calculations? (Remember: The simulator has its own, more complex model of the real transistor, so there should be some small variations.)
- From your simulation, report A_v. How closely does it match your calculations?

PART 2: PROTOTYPING

- Assemble the circuit onto your breadboard using the specified component values and those just calculated. Note that R_{sig} represents the output

resistance of the function generator, and therefore you should *not* include it in your circuit.

PART 3: MEASUREMENTS

- *DC bias point measurement*: Using a digital multimeter, measure the DC voltages of your circuit at the gate (V_G), source (V_S), and drain (V_D) of your transistor.
- *AC measurement*: Using a function generator, apply to your circuit a 10-mV$_{pk-pk}$, 1-kHz sinusoid with no DC component. (*Note*: Some function generators only allow inputs as small as 50 mV$_{pk-pk}$. If this is the case, use that value instead.)
- Using an oscilloscope, generate plots of v_o and v_i vs. t.
- *Output resistance R_o*: Replace R_L with a 1-MΩ resistor and repeat the AC measurement. What is the amplitude of the output waveform? Adjust R_L until you find a value such that the amplitude of the output waveform is approximately 50% of what it was for the 1-MΩ resistor. This new value of R_i is the output resistance R_o. How does it compare to the value you calculated earlier in Step 2? *Hint*: It cannot be greater than the value of R_D.
- *Further exploration*: What happens to the shape of the output signal as you increase the amplitude of the input signal, e.g., to 1 V$_{pk-pk}$? At what input amplitude do you begin to see significant distortion? Can you explain this?
- Using a digital multimeter, measure all resistors to three significant digits.

PART 4: POST-MEASUREMENT EXERCISE

- Calculate the values of V_{SG} and V_{SD} that you obtained in the lab. How do they compare to your pre-lab calculations? Explain any discrepancies.
- Based on the measured values of V_D and V_S and your measured resistor values, what is the real value of I_D based on your lab measurements?
- What is the measured value of A_v? How does it compare to your pre-lab calculations? Explain any discrepancies.
- *Hint*: The single biggest source of variations from your pre-lab simulation results will be due to variations in the transistor's threshold voltage V_{tp}. Remember: Its value will be somewhere within the range indicated on the transistor's datasheet.

PART 5 [OPTIONAL]: EXTRA EXPLORATION

- Instead of tying R_G to ground, try tying it to the drain terminal of the transistor. Repeat the DC bias point measurement and the small-signal gain measurement. What has changed? Do R_D and R_S need to be altered to meet design specifications?

NMOS Common-Source Amplifier with Source Degeneration
(See Section 7.3.4, p. 427 of Sedra/Smith)

OBJECTIVES:

To study an NMOS-based common-source (CS) amplifier with a source-degeneration resistor by:

- Completing the DC and small-signal analysis based on its theoretical behavior.
- Simulating it to compare the results with the paper analysis.
- Implementing it in an experimental setting, taking measurements, and comparing its performance with theoretical and simulated results.
- Qualitatively seeing the impact of transistor-to-transistor variations.

MATERIALS:

- Laboratory setup, including breadboard
- 1 enhancement-type NMOS transistor (e.g., MC14007)
- 3 large (e.g., 47-μF) capacitors
- Several resistors of varying sizes
- Wires

PART 1: DESIGN AND SIMULATION

Consider the circuit shown in Figure L7.3:

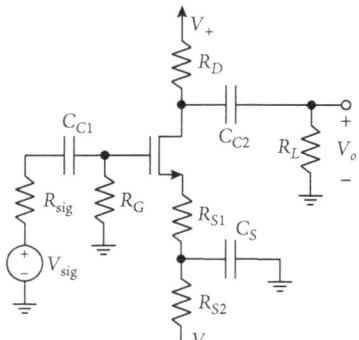

FIGURE L7.3: Common-source amplifier circuit, with source-degeneration resistor R_{S1}, coupling capacitors, and resistor R_G for DC-biasing purposes.

Design the amplifier to achieve a small-signal gain of at least $A_v = -4$ V/V, with $R_{S1} = 220\ \Omega$. Use supplies of $V_+ = -V_- = 15$ V, $R_{sig} = 50\ \Omega$, $R_L = 10$ kΩ, and $R_G = 10$ kΩ, and design the circuit to have $I_D = 1$ mA. Obtain the datasheet for the NMOS transistor that will be used. In your lab book, perform the following.

DC Operating Point Analysis

- Sketch a DC model of the circuit in your lab book, replacing the three large-valued coupling capacitors—C_{C1}, C_{C2}, C_S—by open circuits (for simplicity you may also omit v_{sig}, R_{sig}, and R_L). What is the DC current through R_G?
- Based on the information just given, you have enough information to calculate $V_{OV} = V_{GS} - V_{tn}$. What is its value? What is the value of g_m? What is V_{GS}? Remember: Your actual transistor will have a value of V_{tn} that will vary from its nominal value, which will slightly alter your measurement results!
- Calculate r_o.
- You now have enough information to calculate $R_{S1} + R_{S2}$. Show your calculations. What is R_{S2}?
- Are the values of R_{S1} and R_{S2} available in your kit? Can you achieve the values by combining several resistors? Comment.
- *Note*: At this stage we know neither V_{DS} nor R_D.

AC Analysis

- Sketch a small-signal model of the circuit in your lab book, replacing the transistor with its small-signal model, replacing the capacitors with short circuits (what happens to R_{S2}?), and replacing V_+ with an AC ground. What happens to V_-? Label the gate of the transistor as v_i, i.e., the small-signal voltage at the input.
- What is the ratio of v_i/v_{sig}? How would you approximate it in further calculations?
- Derive an expression for $A_v = v_o/v_i$. What is the value of R_D that produces a small-signal voltage gain of *at least* $A_v = -4$ V/V? Is the value you calculated for R_D available in your kit? Can you achieve this value by combining several resistors? Comment.
- What is the DC voltage at the drain? Does this satisfy the assumption that the transistor should be operating in the saturation region? Explain.

Simulation

- Simulate the performance of your circuit. Use capacitor values $C_{C1} = C_{C2} = C_S = 47\text{-}\mu\text{F}$, and the values of R_{S1}, R_{S2}, and R_D based on your preceding calculations. Use a 10-mV$_{pk-pk}$, 1-kHz sinusoid with no DC component applied at v_{sig}.
- From your simulation, report the DC values of V_{GS}, V_{DS}, and I_D. How closely do they match your calculations? (Remember: The simulator has its own, more complex model of the real transistor, so there should be some small variations.)
- From your simulation, report A_v. How closely does it match your calculations?

PART 2: PROTOTYPING

- Assemble the circuit onto your breadboard using the specified component values and those just calculated. Note that R_{sig} represents the output resistance of the function generator, and therefore you should *not* include it in your circuit.

PART 3: MEASUREMENTS

- *DC bias point measurement*: Using a digital multimeter, measure the DC voltages of your circuit at the gate (V_G), source (V_S), and drain (V_D) of your transistor.
- *AC measurement*: Using a function generator, apply to your circuit a 10-mV$_{pk-pk}$, 1-kHz sinusoid with no DC component. (*Note*: Some function generators only allow inputs as small as 50 mV$_{pk-pk}$. If this is the case, use that value instead.)
- Using an oscilloscope, generate plots of v_o and v_i vs. t.
- *Further exploration 1*: What happens to the shape of the output signal as you increase the amplitude of the input signal, e.g., to 1 V$_{pk-pk}$? At what input amplitude do you begin to see significant distortion? Can you explain this?
- *Further exploration 2*: What happens if you decrease R_{S1} to 200 Ω but keep $R_{S1} + R_{S2}$ constant?
- Using a digital multimeter, measure all resistors to three significant digits.

PART 4: POST-MEASUREMENT EXERCISE

- Calculate the values of V_{GS} and V_{DS} that you obtained in the lab. How do they compare to your pre-lab calculations? Explain any discrepancies.
- Based on the measured values of V_D and V_S and your measured resistor values, what is the real value of I_D based on your lab measurements?
- What is the measured value of A_v? How does it compare to your pre-lab calculations? Explain any discrepancies.
- *Hint*: The single biggest source of variations from your pre-lab simulation results will be due to variations in the transistor's threshold voltage V_{tn}. Remember: Its value will be somewhere within the range indicated on the transistor's datasheet.

PART 5 [OPTIONAL]: EXTRA EXPLORATION

- In your circuit, switch R_{S1} and R_{S2}. Measure your new DC bias point and small-signal voltage gain. What has changed? What remains the same? Can you explain this?

LAB 7.4

PMOS Common-Source Amplifier with Source Degeneration
(See Section 7.3.4, p. 427 of Sedra/Smith)

OBJECTIVES:

To study a PMOS-based common-source (CS) amplifier with a source-degeneration resistor by:

- Completing the DC and small-signal analysis based on its theoretical behavior.
- Simulating it to compare the results with the paper analysis.
- Implementing it in an experimental setting, taking measurements, and comparing its performance with theoretical and simulated results.
- Qualitatively seeing the impact of transistor-to-transistor variations.

MATERIALS:

- Laboratory setup, including breadboard
- 1 enhancement-type PMOS transistor (e.g., MC14007)
- 3 large (e.g., 47–µF) capacitors
- Several resistors of varying sizes
- Wires

PART 1: DESIGN AND SIMULATION

Consider the circuit shown in Figure L7.4:

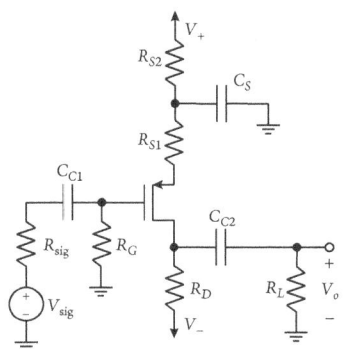

FIGURE L7.4: Common-source amplifier circuit, with source-degeneration resistor R_{S1}, coupling capacitors, and resistor R_G for DC-biasing purposes.

Design the amplifier to achieve a small-signal gain of at least $A_v = -2.5$ V/V, with $R_{S1} = 900$ Ω. Use supplies of $V_+ = -V_- = 15$ V, $R_{sig} = 50$ Ω, $R_L = 10$ kΩ, $R_G = 10$ kΩ, and design the circuit to have $I_D = 1$ mA. Obtain the datasheet for the PMOS transistor that will be used. In your lab book, perform the following.

DC Operating Point Analysis

- Sketch a DC model of the circuit in your lab book, replacing the three large-valued coupling capacitors—C_{C1}, C_{C2}, C_S—by open circuits (for simplicity you may also omit v_{sig}, R_{sig}, and R_L). What is the DC current through R_G?
- Based on the information just given, you have enough information to calculate $V_{OV} = V_{SG} - |V_{tp}|$. What is its value? What is the value of g_m? What is V_{SG}? Remember: Your actual transistor will have a value of V_{tp} that will vary from its nominal value, which will alter your measurement results slightly!
- Calculate r_o.
- You now have enough information to calculate $R_{S1} + R_{S2}$. Show your calculations. What is R_{S2}?
- Are the values of R_{S1} and R_{S2} available in your kit? Can you achieve the values by combining several resistors? Comment.
- *Note*: At this stage we know neither V_{SD} nor R_D.

AC Analysis

- Sketch a small-signal model of the circuit in your lab book, replacing the transistor with its small-signal model, replacing the capacitors with short circuits (what happens to R_{S2}?), and replacing V_- with an AC ground. What happens to V_+? Label the gate of the transistor as v_i, i.e., the small-signal voltage at the input.
- What is the ratio of v_i/v_{sig}? How would you approximate it in further calculations?
- Derive an expression for $A_v = v_o/v_i$. What is the value of R_D that produces a small-signal voltage gain of *at least* $A_v = -2.5$ V/V? Is the value you calculated for R_D available in your kit? Can you achieve this value by combining several resistors? Comment.
- What is the DC voltage at the drain? Does this satisfy the assumption that the transistor should be operating in the saturation region? Explain.

Simulation

- Simulate the performance of your circuit. Use capacitor values $C_{C1} = C_{C2} = C_S = 47$ µF and the values of R_{S1}, R_{S2}, and R_D based on your preceding calculations. Use a 10-mV$_{pk-pk}$, 1-kHz sinusoid with no DC component applied at v_{sig}.
- From your simulation, report the DC values of V_{SG}, V_{SD}, and I_D. How closely do they match your calculations? (Remember: The simulator has its own, more complex model of the real transistor, so there should be some small variations.)
- From your simulation, report A_v. How closely does it match your calculations?

PART 2: PROTOTYPING

- Assemble the circuit onto your breadboard using the specified component values and those just calculated. Note that R_{sig} represents the output resistance of the function generator, and therefore you should *not* include it in your circuit.

PART 3: MEASUREMENTS

- *DC bias point measurement*: Using a digital multimeter, measure the DC voltages of your circuit at the gate (V_G), source (V_S), and drain (V_D) of your transistor.
- *AC measurement*: Using a function generator, apply to your circuit a 10-mV$_{pk-pk}$, 1-kHz sinusoid with no DC component. (Note: Some function generators only allow inputs as small as 50 mV$_{pk-pk}$. If this is the case, use that value instead.)
- Using an oscilloscope, generate plots of v_o and v_i vs. t.
- *Further exploration 1*: What happens to the shape of the output signal as you increase the amplitude of the input signal, e.g., to 1 V$_{pk-pk}$? At what input amplitude do you begin to see significant distortion? Can you explain this?
- *Further exploration 2*: What happens if you decrease R_{S1} to 200 Ω but keep $R_{S1} + R_{S2}$ constant?
- Using a digital multimeter, measure all resistors to three significant digits.

PART 4: POST-MEASUREMENT EXERCISE

- Calculate the values of V_{SG} and V_{SD} that you obtained in the lab. How do they compare to your pre-lab calculations? Explain any discrepancies.
- Based on the measured values of V_D and V_S and your measured resistor values, what is the real value of I_D based on your lab measurements?
- What is the measured value of A_v? How does it compare to your pre-lab calculations? Explain any discrepancies.
- *Hint*: The single biggest source of variations from your pre-lab simulation results will be due to variations in the transistor's threshold voltage V_{tp}. Remember: Its value will be somewhere within the range indicated on the transistor's datasheet.

PART 5 [OPTIONAL]: EXTRA EXPLORATION

- In your circuit, switch R_{S1} and R_{S2}. Measure your new DC bias point and small-signal voltage gain. What has changed? What remains the same? Can you explain this?

NMOS Common-Gate Amplifier
(See Section 7.3.5, p. 434 of Sedra/Smith)

OBJECTIVES:

To study an NMOS-based common-gate (CG) amplifier by:

- Completing the DC and small-signal analysis based on its theoretical behavior.
- Simulating it to compare the results with the paper analysis.
- Implementing it in an experimental setting, taking measurements, and comparing its performance with theoretical and simulated results.
- Qualitatively seeing the impact of transistor-to-transistor variations.

MATERIALS:

- Laboratory setup, including breadboard
- 1 enhancement-type NMOS transistor (e.g., MC14007)
- 2 large (e.g., 47-μF) capacitors
- Several resistors of varying sizes
- Wires

PART 1: DESIGN AND SIMULATION

Consider the circuit shown in Figure L7.5:

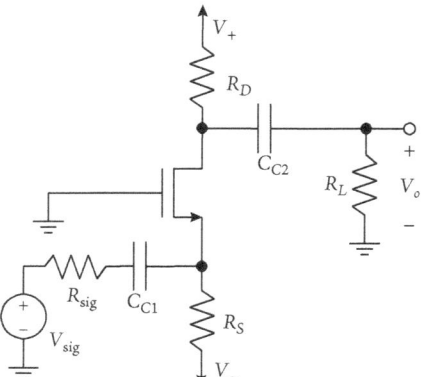

FIGURE L7.5: Common-gate amplifier circuit, with coupling capacitors.

Design the amplifier to achieve a small-signal gain of at least $A_v = 5$ V/V. Use supplies of $V_+ = -V_- = 15$ V, $R_{sig} = 50$ Ω, and $R_L = 10$ kΩ, and design the circuit to have $I_D = 1$ mA. Obtain the datasheet for the NMOS transistor that will be used. In your lab book, perform the following.

DC Operating Point Analysis

- Sketch a DC model of the circuit in your lab book, replacing the "large-valued" coupling capacitors, C_{C1} and C_{C2}, by open circuits (for simplicity you may also omit v_{sig}, R_{sig}, and R_L).
- Based on the information just given, you have enough information to calculate $V_{OV} = V_{GS} - V_{tn}$. What is its value? What is the value of g_m? What is V_{GS}? Remember: Your actual transistor will have a value of V_{tn} that will vary from its nominal value, which will alter your measurement results slightly!
- Calculate r_o.
- You now have enough information to calculate R_S. Show your calculations. Is the value you calculate for R_S available in your kit? Can you achieve this value by combining several resistors? Comment.
- *Note*: At this stage we know neither V_{DS} nor R_D.

AC Analysis

- Sketch a small-signal model of the circuit in your lab book, replacing the transistor with its small-signal model, replacing the capacitors with short circuits, and replacing V_+ and V_- with an AC ground. Label the source of the transistor as v_i, i.e., the small-signal voltage at the input.
- What is the ratio of v_i/v_{sig}? How would you approximate it in further calculations?
- Derive an expression for $A_v = v_o/v_i$. What is the value of R_D that produces a small-signal voltage gain of *at least* $A_v = 5$ V/V? Is the value you calculate for R_D available in your kit? Can you achieve this value by combining several resistors? Comment.
- What is the DC voltage at the drain? Does this satisfy the assumption that the transistor should be operating in the saturation region? Explain.

Simulation

- Simulate the performance of your circuit. Use capacitor values $C_{C1} = C_{C2} = 47\,\mu$F and the values of R_S and R_D based on your preceding calculations. Use a 10-mV$_{pk-pk}$, 1-kHz sinusoid with no DC component applied at v_{sig}.
- From your simulation, report the DC values of V_{GS}, V_{DS}, and I_D. How closely do they match your calculations? (Remember: The simulator has its own, more complex model of the real transistor, so there should be some small variations.)
- From your simulation, report A_v. How closely does it match your calculations?

PART 2: PROTOTYPING

- Assemble the circuit onto your breadboard using the specified component values and those just calculated. Note that R_{sig} represents the output

resistance of the function generator, and therefore you should *not* include it in your circuit.

PART 3: MEASUREMENTS

- *DC bias point measurement*: Using a digital multimeter, measure the DC voltages of your circuit at the gate (V_G), source (V_S), and drain (V_D) of your transistor.
- *AC measurement*: Using a function generator, apply a 10-mV$_{pk-pk}$, 1-kHz sinusoid with no DC component to your circuit. (*Note*: Some function generators only allow inputs as small as 50 mV$_{pk-pk}$. If this is the case, use that value instead.)
- Using an oscilloscope, generate plots of v_o and v_i vs. t.
- *Further exploration*: What happens to the shape of the output signal as you increase the amplitude of the input signal, e.g., to 2 V$_{pk-pk}$? At what input amplitude do you begin to see significant distortion?
- Using a digital multimeter, measure all resistors to three significant digits.

PART 4: POST-MEASUREMENT EXERCISE

- Calculate the values of V_{GS} and V_{DS} that you obtained in the lab. How do they compare to your pre-lab calculations? Explain any discrepancies.
- Based on the measured values of V_D and V_S and your measured resistor values, what is the real value of I_D based on your lab measurements?
- What is the measured value of A_v? How does it compare to your pre-lab calculations? Explain any discrepancies.
- *Hint*: The single biggest source of variations from your pre-lab simulation results will be due to variations in the transistor's threshold voltage V_{tn}. Remember: Its value will be somewhere within the range indicated on the transistor's datasheet.

PART 5 [OPTIONAL]: EXTRA EXPLORATION

- Using the function generator, add a 1-V DC component to the input v_{sig} and repeat the measurements. Do you still get the same DC operating point and voltage gain A_v? Why or why not?

PMOS Common-Gate Amplifier
(See Section 7.3.5, p. 434 of Sedra/Smith)

OBJECTIVES:

To study a PMOS-based common-gate (CG) amplifier by:

*   Completing the DC and small-signal analysis based on its theoretical behavior.
*   Simulating it to compare the results with the paper analysis.
*   Implementing it in an experimental setting, taking measurements, and comparing its performance with theoretical and simulated results.
*   Qualitatively seeing the impact of transistor-to-transistor variations.

MATERIALS:

*   Laboratory setup, including breadboard
*   1 enhancement-type PMOS transistor (e.g., MC14007)
*   2 large (e.g., 47-μF) capacitors
*   Several resistors of varying sizes
*   Wires

PART 1: DESIGN AND SIMULATION

Consider the circuit shown in Figure L7.6:

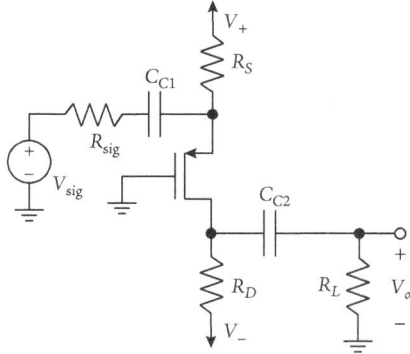

FIGURE L7.6: Common-gate amplifier circuit, with coupling capacitors.

Design the amplifier to achieve a small-signal gain of at least $A_v = 5$ V/V. Use supplies of $V_+ = -V_- = 15$ V, $R_{sig} = 50\ \Omega$, $R_L = 10$ kΩ, and design the circuit to have $I_D = 1$ mA. Obtain the datasheet for the PMOS transistor that will be used. In your lab book, perform the following.

DC Operating Point Analysis

- Sketch a DC model of the circuit in your lab book, replacing the "large-valued" coupling capacitors, C_{C1} and C_{C2}, by open circuits (for simplicity you may also omit v_{sig}, R_{sig}, and R_L).
- Based on the information just given, you have enough information to calculate $V_{OV} = V_{SG} - |V_{tp}|$. What is its value? What is the value of g_m? What is V_{SG}? Remember: Your actual transistor will have a value of V_{tp} that will vary from its nominal value, which will alter your measurement results slightly!
- Calculate r_o.
- You now have enough information to calculate R_S. Show your calculations. Is the value you calculate for R_S available in your kit? Can you achieve this value by combining several resistors? Comment.
- *Note*: At this stage we know neither V_{SD} nor R_D.

AC Analysis

- Sketch a small-signal model of the circuit in your lab book, replacing the transistor with its small-signal model, replacing the capacitors with short circuits, and replacing V_+ and V_- with an AC ground. Label the source of the transistor as v_i, i.e., the small-signal voltage at the input.
- What is the ratio of v_i/v_{sig}? How would you approximate it in further calculations?
- Derive an expression for $A_v = v_o/v_i$. What is the value of R_D that produces a small-signal voltage gain of *at least* $A_v = 5$ V/V? Is the value you calculate for R_D available in your kit? Can you achieve this value by combining several resistors? Comment.
- What is the DC voltage at the drain? Does this satisfy the assumption that the transistor should be operating in the saturation region? Explain.

Simulation

- Simulate the performance of your circuit. Use capacitor values $C_{C1} = C_{C2} = 47\ \mu$F and the values of R_S and R_D based on your preceding calculations. Use a 10-mV$_{pk-pk}$, 1-kHz sinusoid with no DC component applied at v_{sig}.
- From your simulation, report the DC values of V_{SG}, V_{SD}, and I_D. How closely do they match your calculations? (Remember: The simulator has its own, more complex model of the real transistor, so there should be some small variations.)
- From your simulation, report A_v. How closely does it match your calculations?

PART 2: PROTOTYPING

- Assemble the circuit onto your breadboard using the component values specified and calculated earlier. Note that R_{sig} represents the output resistance of the function generator, and therefore you should *not* include it in your circuit.

PART 3: MEASUREMENTS

- *DC bias point measurement*: Using a digital multimeter, measure the DC voltages of your circuit at the gate (V_G), source (V_S), and drain (V_D) of your transistor.
- *AC measurement*: Using a function generator, apply a 10-mV$_{pk-pk}$, 1-kHz sinusoid with no DC component to your circuit. (*Note*: Some function generators only allow inputs as small as 50 mV$_{pk-pk}$. If this is the case, use that value instead.)
- Using an oscilloscope, generate plots of v_o and v_i vs. t.
- *Further exploration*: What happens to the shape of the output signal as you increase the amplitude of the input signal, e.g., to 2 V$_{pk-pk}$? At what input amplitude do you begin to see significant distortion?
- Using a digital multimeter, measure all resistors to three significant digits.

PART 4: POST-MEASUREMENT EXERCISE

- Calculate the values of V_{SG} and V_{SD} that you obtained in the lab. How do they compare to your pre-lab calculations? Explain any discrepancies.
- Based on the measured values of V_D and V_S and your measured resistor values, what is the real value of I_D based on your lab measurements?
- What is the measured value of A_v? How does it compare to your pre-lab calculations? Explain any discrepancies.
- *Hint*: The single biggest source of variations from your pre-lab simulation results will be due to variations in the transistor's threshold voltage V_{tp}. Remember: Its value will be somewhere within the range indicated on the transistor's datasheet.

PART 5 [OPTIONAL]: EXTRA EXPLORATION

- Using the function generator, add a 1-V DC component to the input v_{sig} and repeat the measurements. Do you still get the same DC operating point and voltage gain A_v? Why or why not?

NMOS Source Follower
(See Section 7.3.6, p. 437 of Sedra/Smith)

OBJECTIVES:

To study an NMOS-based source follower by:

- Completing the DC and small-signal analysis based on its theoretical behavior.
- Simulating it to compare the results with the paper analysis.
- Implementing it in an experimental setting, taking measurements, and comparing its performance with theoretical and simulated results.
- Qualitatively seeing the impact of transistor-to-transistor variations.

MATERIALS:

- Laboratory setup, including breadboard
- 1 enhancement-type NMOS transistor (e.g., MC14007)
- 3 large (e.g., 47-μF) capacitors
- Several resistors of varying sizes
- Wires

PART 1: DESIGN AND SIMULATION

Consider the circuit shown in Figure L7.7:

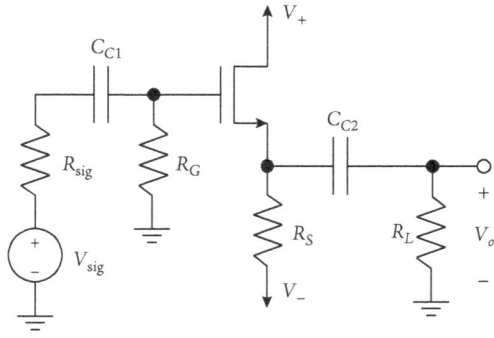

FIGURE L7.7: Source follower circuit, with coupling capacitors, and resistor R_G for DC-biasing purposes.

- Design the amplifier such that $I_D = 1$ mA and $A_v = 0.8$ V/V. Use supplies of $V_+ = -V_- = 15$ V, $R_{sig} = 50\ \Omega$, and $R_G = 10$ kΩ. What is the minimum value of R_L that satisfies the requirements? Obtain the datasheet for the NMOS transistor that will be used. In your lab book, perform the following.

DC Operating Point Analysis

- Sketch a DC model of the circuit in your lab book, replacing the large-valued coupling capacitors C_{C1} and C_{C2} by open circuits (for simplicity you may also omit v_{sig}, R_{sig}, and R_L). What is the DC current through R_G?
- Based on the required value of I_D, what is $V_{OV} = V_{GS} - V_{tn}$? What value of R_S must you use?

AC Analysis

- Sketch a small-signal model of the circuit in your lab book, replacing the transistor with its small-signal model (try a T model, ignoring r_o), replacing the capacitors with short circuits, and replacing V_+ and V_- with an AC ground. Label the gate of the transistor as v_i, i.e., the small-signal voltage at the input.
- What is the ratio of v_i/v_{sig}? How would you approximate it in further calculations?
- Derive an expression for $A_v = v_o/v_i$.
- What is the value of g_m? What is A_v?
- What is the minimum value of R_L that satisfies the design requirements?
- Calculate the output resistance of your amplifier.

Simulation

- Simulate the performance of your circuit. Use capacitor values $C_{C1} = C_{C2} = 47$ µF and the value of R_S based on your preceding calculations. Use a 10-mV$_{pk-pk}$, 1-kHz sinusoid with no DC component applied at v_{sig}.
- From your simulation, report the DC values of V_{GS}, V_{DS}, and I_D. How closely do they match your calculations? (Remember: The simulator has its own, more complex model of the real transistor, so there should be some small variations.)
- From your simulation, report A_v. How closely does it match your calculations?

PART 2: PROTOTYPING

- Assemble the circuit onto your breadboard using the specified component values and those just calculated. Note that R_{sig} represents the output resistance of the function generator, and therefore you should *not* include it in your circuit.

PART 3: MEASUREMENTS

- *DC bias point measurement*: Using a digital multimeter, measure the DC voltages of your circuit at the gate (V_G) and source (V_S) of your transistor.

- *AC measurement*: Using a function generator, apply a 10-mV$_{pk-pk}$, 1-kHz sinusoid with no DC component to your circuit. (*Note*: Some function generators only allow inputs as small as 50 mV$_{pk-pk}$. If this is the case, use that value instead.)
- Using an oscilloscope, generate plots of v_o and v_i vs. *t*.
- Using a digital multimeter, measure all resistors to three significant digits.

PART 4: POST-MEASUREMENT EXERCISE

- Calculate the values of V_{GS} and V_{DS} that you obtained in the lab. How do they compare to your pre-lab calculations? Explain any discrepancies.
- Based on the measured values of V_D and V_S and your measured resistor values, what is the real value of I_D based on your lab measurements?
- What is the measured value of A_v? How does it compare to your pre-lab calculations? Explain any discrepancies.
- What would happen if you used the function generator with 50-Ω output resistance to drive your load resistor directly? What gain would you get? What would happen if the output resistance of the function generation was changed from 50 Ω to 5 kΩ? What do you conclude? Recall the value of output resistance you calculated earlier.
- *Hint*: The single biggest source of variations from your pre-lab simulation results will be due to variations in the transistor's threshold voltage V_{tn}. Remember: Its value will be somewhere within the range indicated on the transistor's datasheet.

PART 5 [OPTIONAL]: EXTRA EXPLORATION

- Add a 500-Ω resistor between the function generator output and capacitor C_{C1}. How does the gain of your circuit change? Can you explain this?

PMOS Source Follower
(See Section 7.3.6, p. 437 of Sedra/Smith)

OBJECTIVES:

To study a PMOS-based source follower by:

- Completing the DC and small-signal analysis based on its theoretical behavior.
- Simulating it to compare the results with the paper analysis.
- Implementing it in an experimental setting, taking measurements, and comparing its performance with theoretical and simulated results.
- Qualitatively seeing the impact of transistor-to-transistor variations.

MATERIALS:

- Laboratory setup, including breadboard
- 1 enhancement-type PMOS transistor (e.g., MC14007)
- 3 large (e.g., 47-μF) capacitors
- Several resistors of varying sizes
- Wires

PART 1: DESIGN AND SIMULATION

Consider the circuit shown in Figure L7.8:

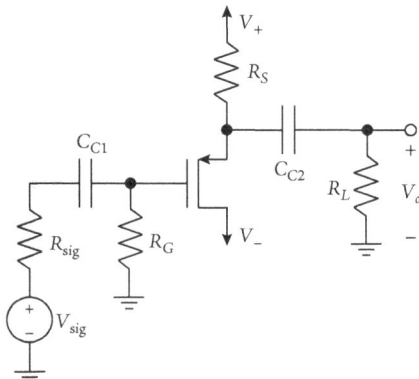

FIGURE L7.8: Source follower circuit with coupling capacitors, and resistor R_G for DC-biasing purposes.

Design the amplifier such that $I_D = 2$ mA. Use supplies of $V_+ = -V_- = 15$ V, $R_{\text{sig}} = 50 \, \Omega$, and $R_G = 10 \, \text{k}\Omega$. What is the minimum value of R_L that satisfies the requirements? Obtain the datasheet for the PMOS transistor that will be used. In your lab book, perform the following.

DC Operating Point Analysis

- Sketch a DC model of the circuit in your lab book, replacing the large-valued coupling capacitors, C_{C1} and C_{C2}, by open circuits (for simplicity you may also omit v_{sig}, R_{sig}, and R_L). What is the DC current through R_G?
- Based on the required value of I_D, what is $V_{OV} = V_{SG} - |V_{tp}|$? What value of R_S must you use?

AC Analysis

- Sketch a small-signal model of the circuit in your lab book, replacing the transistor with its small-signal model (try a T model, ignoring r_o), replacing the capacitors with short circuits, and replacing V_+ and V_- with an AC ground. Label the gate of the transistor as v_i, i.e., the small-signal voltage at the input.
- What is the ratio of v_i/v_{sig}? How would you approximate it in further calculations?
- Derive an expression for $A_v = v_o/v_i$.
- What is the value of g_m? What is A_v?
- What is the minimum value of R_L that satisfies the design requirements?
- Calculate the output resistance of your amplifier.

Simulation

- Simulate the performance of your circuit. Use capacitor values $C_{C1} = C_{C2} = 47$ μF and the value of R_S based on your preceding calculations. Use a 10-mV$_{\text{pk-pk}}$, 1-kHz sinusoid with no DC component applied at v_{sig}.
- From your simulation, report the DC values of V_{SG}, V_{SD}, and I_D. How closely do they match your calculations? (Remember: The simulator has its own, more complex model of the real transistor, so there should be some small variations.)
- From your simulation, report A_v. How closely does it match your calculations?

PART 2: PROTOTYPING

- Assemble the circuit onto your breadboard using the specified component values and those just calculated. Note that R_{sig} represents the output resistance of the function generator, and therefore you should *not* include it in your circuit.

PART 3: MEASUREMENTS

- *DC bias point measurement*: Using a digital multimeter, measure the DC voltages of your circuit at the gate (V_G) and source (V_S) of your transistor.

- *AC measurement*: Using a function generator, apply a 10-mV$_{pk-pk}$, 1-kHz sinusoid with no DC component to your circuit. (*Note*: Some function generators only allow inputs as small as 50 mV$_{pk-pk}$. If this is the case, use that value instead.)
- Using an oscilloscope, generate plots of v_o and v_i vs. t.
- Using a digital multimeter, measure all resistors to three significant digits.

PART 4: POST-MEASUREMENT EXERCISE

- Calculate the values of V_{SG} and V_{SD} that you obtained in the lab. How do they compare to your pre-lab calculations? Explain any discrepancies.
- Based on the measured values of V_D and V_S, and your measured resistor values, what is the real value of I_D based on your lab measurements?
- What is the measured value of A_v? How does it compare to your pre-lab calculations? Explain any discrepancies.
- What would happen if you used the function generator with 50-Ω output resistance to directly drive your load resistor? What gain would you get? What would happen if the output resistance of the function generation was changed from 50 Ω to 5 kΩ? What do you conclude? Recall the value of output resistance you calculated earlier.
- *Hint*: The single biggest source of variations from your pre-lab simulation results will be due to variations in the transistor's threshold voltage V_{tp}. Remember: Its value will be somewhere within the range indicated on the transistor's datasheet.

PART 5 [OPTIONAL]: EXTRA EXPLORATION

- Add a 500-Ω resistor between the function generator output and capacitor C_{C1}. How does the gain of your circuit change? Can you explain this?

NPN Common-Emitter Amplifier
(See Section 7.5.2, p. 464 of Sedra/Smith)

OBJECTIVES:

To study an NPN-based common-emitter (CE) amplifier by:

- Completing the DC and small-signal analysis based on its theoretical behavior.
- Simulating it to compare the results with the paper analysis.
- Implementing it in an experimental setting, taking measurements, and comparing its performance with theoretical and simulated results.
- Measuring its output resistance.
- Qualitatively seeing the impact of transistor-to-transistor variations.

MATERIALS:

- Laboratory setup, including breadboard
- 1 NPN-type bipolar transistor (e.g., NTE2321)
- 3 "large" (e.g., 47-μF) capacitors
- Several resistors of varying sizes
- Wires

PART 1: DESIGN AND SIMULATION

Consider the circuit shown in Figure L7.9:

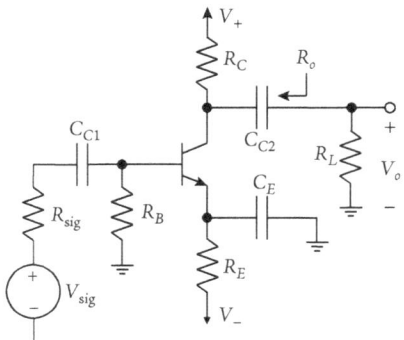

FIGURE L7.9: Common-emitter amplifier circuit, with coupling capacitors, and resistor R_B for DC-biasing purposes. Based on Fig. 7.58 p. 465 S&S.

Design the amplifier to achieve a small-signal gain of at least $A_v = -200$ V/V. Use supplies of $V_+ = -V_- = 15$ V, $R_{sig} = 50\ \Omega$, $R_L = 10\ k\Omega$, and $R_B = 10\ k\Omega$, and design the circuit to have $I_C = 1$ mA. Although there will be variations from transistor to transistor, you may assume a value of β of 100 in your calculations. Obtain the datasheet for the NPN transistor that will be used. In your lab book, perform the following:

DC Operating Point Analysis

- Sketch a DC model of the circuit in your lab book, replacing the three "large-valued" coupling capacitors—C_{C1}, C_{C2}, and C_E—by open circuits (for simplicity you may also omit v_{sig}, R_{sig}, and R_L).
- What are the values of I_B and I_E? What is the value of V_B?
- Determine a value of R_E that produces a base-emitter voltage drop of 0.7 V. What is V_E?
- Is the value of R_E available in your kit? Can you achieve this value by combining several resistors? Comment.
- *Note*: At this stage we know neither V_{CE} nor R_C.

AC Analysis

- Sketch a small-signal model of the circuit in your lab book, replacing the transistor with its small-signal model (V_A is large, so you may ignore r_o), replacing the capacitors with short circuits (what happens to R_E?), and replacing V_+ with an AC ground. What happens to V_-? Label the base of the transistor as v_i, i.e., the small-signal voltage at the input. What are the values of g_m and r_π?
- What is the ratio of v_i/v_{sig}? Can you approximate it?
- Derive an expression for $A_v = v_o/v_i$. What is the value of R_C that produces a small-signal voltage gain of *at least* $A_v = -200$ V/V? Is the value you calculate for R_C available in your kit? Can you achieve this value by combining several resistors? Comment.
- What is the DC voltage at the collector? Does this satisfy the assumption that the transistor should be operating in the active region? Explain.
- What is the output resistance, R_o?

Simulation

- Simulate the performance of your circuit. Use capacitor values $C_{C1} = C_{C2} = C_E = 47\ \mu F$ and the values of R_E and R_C based on your preceding calculations. Use a 10-mV$_{pk-pk}$, 1-kHz sinusoid with no DC component applied at v_{sig}.
- From your simulation, report the DC values of V_{BE}, V_{CE}, I_B, I_C, and I_E. How closely do they match your calculations?
- From your simulation, report A_v. How closely does it match your calculations?

PART 2: PROTOTYPING

- Assemble the circuit onto your breadboard using the specified component values and those just calculated. Note that R_{sig} represents the output

resistance of the function generator, and therefore you should *not* include it in your circuit.

PART 3: MEASUREMENTS

- *DC bias point measurement*: Using a digital multimeter, measure the DC voltages of your circuit at the base (V_B), emitter (V_E), and collector (V_C) of your transistor.
- *AC measurement*: Using a function generator, apply a 10-mV$_{pk–pk}$, 1-kHz sinusoid with no DC component to your circuit. (*Note*: Some function generators only allow inputs as small as 50 mV$_{pk–pk}$. If this is the case, use that value instead, but you may expect some distortion in the output waveform.)
- Using an oscilloscope, generate plots of v_o and v_i vs. *t*.
- *Output resistance R_o*: Replace R_L with a 1-MΩ resistor and repeat the AC measurement. What is the amplitude of the output waveform? Adjust R_L until you find a value such that the amplitude of the output waveform is approximately 50% of what it was for the 1-MΩ resistor. This new value of R_L is the output resistance R_o. How does it compare to the value you calculated earlier in Step 2? *Hint*: It cannot be greater than the value of R_C.
- *Further exploration*: What happens to the shape of the output signal as you increase the amplitude of the input signal, e.g., to 1 V$_{pk–pk}$? At what input amplitude do you begin to see significant distortion?
- Using a digital multimeter, measure all resistors to three significant digits.

PART 4: POST-MEASUREMENT EXERCISE

- Calculate the values of V_{BE} and V_{CE} that you obtained in the lab. How do they compare to your pre-lab calculations? Explain any discrepancies.
- Based on the measured values of V_B, V_C, and V_E and your measured resistor values, what are the real values of all currents based on your lab measurements? How do they compare to your pre-lab calculations? Based on the measured values of currents, what is the actual value of β for your transistor?
- What is the measured value of A_v? How does it compare to your pre-lab calculations? Explain any discrepancies.
- *Hint*: The single biggest source of variations from your pre-lab simulation results will be due to variations in β.

PART 5 [OPTIONAL]: EXTRA EXPLORATION

- Instead of tying R_B to ground, try tying it to the collector terminal of the transistor. Repeat the DC bias point measurement and the small-signal gain measurement. What has changed? Do R_C and R_E need to be altered to meet design specifications?

PNP Common-Emitter Amplifier
(See Section 7.5.2, p. 464 of Sedra/Smith)

OBJECTIVES:

To study a PNP-based common-emitter (CE) amplifier by:

- Completing the DC and small-signal analysis based on its theoretical behavior.
- Simulating it to compare the results with the paper analysis.
- Implementing it in an experimental setting, taking measurements, and comparing its performance with theoretical and simulated results.
- Measuring its output resistance.
- Qualitatively seeing the impact of transistor-to-transistor variations.

MATERIALS:

- Laboratory setup, including breadboard
- 1 PNP-type bipolar transistor (e.g., NTE2322)
- 3 "large" (e.g., 47-μF) capacitors
- Several resistors of varying sizes
- Wires

PART 1: DESIGN AND SIMULATION

Consider the circuit shown in Figure L7.10:

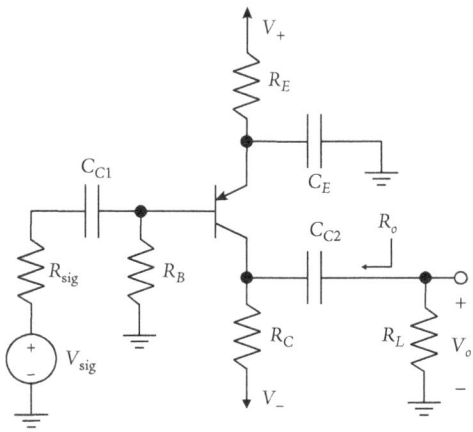

FIGURE L7.10: Common-emitter amplifier circuit, with coupling capacitors, and resistor R_B for DC-biasing purposes. Based on Fig. 7.58 p. 465 S&S.

Design the amplifier to achieve a small-signal gain of at least $A_v = -200$ V/V. Use supplies of $V_+ = -V_- = 15$ V, $R_{sig} = 50\ \Omega$, $R_L = 10$ kΩ, $R_B = 10$ kΩ, and design the circuit to have $I_C = 1$ mA. Although there will be variations from transistor to transistor, you may assume a value of β of 100 in your calculations. Obtain the datasheet for the PNP transistor that will be used. In your lab book, perform the following:

DC Operating Point Analysis

- Sketch a DC model of the circuit in your lab book, replacing the three "large-valued" coupling capacitors—C_{C1}, C_{C2}, and C_E—by open circuits (for simplicity you may also omit v_{sig}, R_{sig}, and R_L).
- What are the values of I_B and I_E? What is the value of V_B?
- Determine a value of R_E that produces an emitter-base voltage drop of approximately 0.7 V. What is V_E?
- Is the value of R_E available in your kit? Can you achieve this value by combining several resistors? Comment.
- Note: At this stage we know neither V_{EC} nor R_C.

AC Analysis

- Sketch a small-signal model of the circuit in your lab book, replacing the transistor with its small-signal model (V_A is large so you may ignore r_o), replacing the capacitors with short circuits (What happens to R_E?), and replacing V_- with an AC ground. What happens to V_+? Label the base of the transistor as v_i, i.e., the small-signal voltage at the input. What are the values of g_m and r_π?
- What is the ratio of v_i/v_{sig}? Can you approximate it?
- Derive an expression for $A_v = v_o/v_i$. What is the value of R_C that produces a small-signal voltage gain of *at least* $A_v = -200$ V/V? Is the value you calculate for R_C available in your kit? Can you achieve this value by combining several resistors? Comment.
- What is the DC voltage at the collector? Does this satisfy the assumption that the transistor should be operating in the active region? Explain.
- What is the output resistance, R_o?

Simulation

- Simulate the performance of your circuit. Use capacitor values $C_{C1} = C_{C2} = C_E = 47\,\mu$F and the values of R_E and R_C based on your preceding calculations. Use a 10-mV$_{pk-pk}$, 1-kHz sinusoid with no DC component applied at v_{sig}.
- From your simulation, report the DC values of V_{EB}, V_{EC}, I_B, I_C, and I_E. How closely do they match your calculations?
- From your simulation, report A_v. How closely does it match your calculations?

PART 2: PROTOTYPING

- Assemble the circuit onto your breadboard using the specified component values and those calculated earlier. Note that R_{sig} represents the output resistance of the function generator, and therefore you should *not* include it in your circuit.

PART 3: MEASUREMENTS

- *DC bias point measurement*: Using a digital multimeter, measure the DC voltages of your circuit at the base (V_B), emitter (V_E), and collector (V_C) of your transistor.
- *AC measurement*: Using a function generator, apply a 10-mV$_{pk-pk}$, 1-kHz sinusoid with no DC component to your circuit. (*Note*: Some function generators only allow inputs as small as 50 mV$_{pk-pk}$. If this is the case, use that value instead, but you may expect some distortion in the output waveform.)
- Using an oscilloscope, generate plots of v_o and v_i vs. t.
- *Output resistance R_o*: Replace R_L with a 1-MΩ resistor and repeat the AC measurement. What is the amplitude of the output waveform? Adjust R_L until you find a value such that the amplitude of the output waveform is approximately 50% of what it was for the 1-MΩ resistor. This new value of R_L is the output resistance R_o. How does it compare to the value you calculated earlier in Step 2? *Hint*: It cannot be greater than the value of R_C.
- *Further exploration*: What happens to the shape of the output signal as you increase the amplitude of the input signal, e.g., to 1V$_{pk-pk}$? At what input amplitude do you begin to see significant distortion?
- Using a digital multimeter, measure all resistors to three significant digits.

PART 4: POST-MEASUREMENT EXERCISE

- Calculate the values of V_{EB} and V_{EC} that you obtained in the lab. How do they compare to your pre-lab calculations? Explain any discrepancies.
- Based on the measured values of V_B, V_C, and V_E and your measured resistor values, what are the real values of all currents based on your lab measurements? How do they compare to your pre-lab calculations? Based on the measured values of currents, what is the actual value of β for your transistor?
- What is the measured value of A_v? How does it compare to your pre-lab calculations? Explain any discrepancies.
- *Hint*: The single biggest source of variations from your pre-lab simulation results will be due to variations in β.

PART 5 [OPTIONAL]: EXTRA EXPLORATION

- Instead of tying R_B to ground, try tying it to the collector terminal of the transistor. Repeat the DC bias point measurement and the small-signal gain measurement. What has changed? Do R_C and R_E need to be altered to meet design specifications?

NPN Common-Emitter Amplifier with Emitter Degeneration
(See Section 7.5.3, p. 466 of Sedra/Smith)

OBJECTIVES:

To study an NPN-based common-emitter (CE) amplifier with an emitter-degeneration resistor by:

- Completing the DC and small-signal analysis based on its theoretical behavior.
- Simulating it to compare the results with the paper analysis.
- Implementing it in an experimental setting, taking measurements, and comparing its performance with theoretical and simulated results.
- Qualitatively seeing the impact of transistor-to-transistor variations.

MATERIALS:

- Laboratory setup, including breadboard
- 1 NPN transistor (e.g., NTE2321)
- 3 large (e.g., 47-μF) capacitors
- Several resistors of varying sizes
- Wires

PART 1: DESIGN AND SIMULATION

Consider the circuit shown in Figure L7.11:

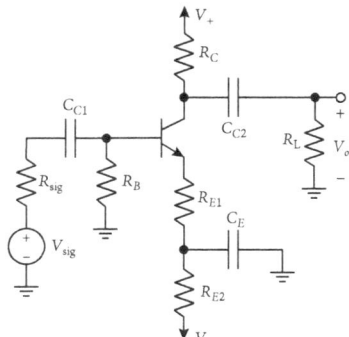

FIGURE L7.11: Common-emitter amplifier circuit, with emitter-degeneration resistor R_{E1}, coupling capacitors, and resistor R_B for DC-biasing purposes. Based on Fig. 7.59 p. 467 S&S.

Design the amplifier to achieve a small-signal gain of at least $A_v = -2.2$ V/V, with $R_{E1} = 2.3$ kΩ. Use supplies of $V_+ = -V_- = 15$ V, $R_{sig} = 50$ Ω, $R_L = 10$ kΩ, and $R_B = 10$ kΩ, and design the circuit to have $I_C = 1$ mA. Obtain the datasheet for the NPN transistor that will be used. In your lab book, perform the following.

DC Operating Point Analysis

- Sketch a DC model of the circuit in your lab book, replacing the three "large-valued" coupling capacitors—C_{C1}, C_{C2}, and C_E—by open circuits (for simplicity you may also omit v_{sig}, R_{sig}, and R_L).
- What are the values of I_B and I_E? What is the value of V_B?
- Determine a value of $R_{E1} + R_{E2}$ that produces a base-emitter voltage drop of 0.7 V. What is V_E? What is R_{E2}?
- Are the values of R_{E1} and R_{E2} available in your kit? Can you achieve these values by combining several resistors? Comment.
- *Note*: At this stage we know neither V_{CE} nor R_C.

AC Analysis

- Sketch a small-signal model of the circuit in your lab book, replacing the transistor with its small-signal model (V_A is large, so you may ignore r_o), replacing the capacitors with short circuits (What happens to R_{E2}?), and replacing V_+ with an AC ground. What happens to V_-? Label the base of the transistor as v_i, i.e., the small-signal voltage at the input. What are the values of g_m and r_π?
- What is the ratio of v_i/v_{sig}? Can you approximate it?
- Derive an expression for $A_v = v_o/v_i$. What is the value of R_C that produces a small-signal voltage gain of *at least* $A_v = -2.2$ V/V? Is the value you calculate for R_C available in your kit? Can you achieve this value by combining several resistors? Comment.
- What is the DC voltage at the collector? Does this satisfy the assumption that the transistor should be operating in the active region? Explain.

Simulation

- Simulate the performance of your circuit. Use capacitor values $C_{C1} = C_{C2} = C_E = 47$ μF and the values of R_{E1}, R_{E2}, and R_C based on your preceding calculations. Use a 10-mV$_{pk-pk}$, 1-kHz sinusoid with no DC component applied at v_{sig}.
- From your simulation, report the DC values of V_{BE}, V_{CE}, I_B, I_C, and I_E. How closely do they match your calculations?
- From your simulation, report A_v. How closely does it match your calculations?

PART 2: PROTOTYPING

- Assemble the circuit onto your breadboard using the specified component values and those just calculated. Note that R_{sig} represents the output resistance of the function generator, and therefore you should *not* include it in your circuit.

PART 3: MEASUREMENTS

- *DC bias point measurement*: Using a digital multimeter, measure the DC voltages of your circuit at the base (V_B), emitter (V_E), and collector (V_C) of your transistor.
- *AC measurement*: Using a function generator, apply a 10-mV$_{pk-pk}$, 1-kHz sinusoid with no DC component to your circuit. (*Note*: Some function generators only allow inputs as small as 50 mV$_{pk-pk}$. If this is the case, use that value instead.)
- Using an oscilloscope, generate plots of v_o and v_i vs. t.
- *Further exploration 1*: What happens to the shape of the output signal as you increase the amplitude of the input signal, e.g., to 1 V$_{pk-pk}$? At what input amplitude do you begin to see significant distortion?
- *Further exploration 2*: What happens if you decrease R_{E1} to 500 Ω but keep $R_{E1} + R_{E2}$ constant?
- Using a digital multimeter, measure all resistors to three significant digits.

PART 4: POST-MEASUREMENT EXERCISE

- Calculate the values of V_{BE} and V_{CE} that you obtained in the lab. How do they compare to your pre-lab calculations? Explain any discrepancies.
- Based on the measured values of V_B, V_C, and V_E and your measured resistor values, what are the real values of all currents based on your lab measurements? How do they compare to your pre-lab calculations? Based on the measured values of currents, what is the actual value of β for your transistor?
- What is the measured value of A_v? How does it compare to your pre-lab calculations? Explain any discrepancies.
- *Hint*: The single biggest source of variations from your pre-lab simulation results will be due to variations in β.

PART 5 [OPTIONAL]: EXTRA EXPLORATION

- In your circuit, switch R_{E1} and R_{E2}. Measure your new DC bias point and small-signal voltage gain. What has changed? What remains the same? Can you explain this?

PNP Common-Emitter Amplifier with Emitter Degeneration
(See Section 7.5.3, p. 466 of Sedra/Smith)

OBJECTIVES:

To study a PNP-based common-emitter (CE) amplifier with an emitter-degeneration resistor by:

- Completing the DC and small-signal analysis based on its theoretical behavior.
- Simulating it to compare the results with the paper analysis.
- Implementing it in an experimental setting, taking measurements, and comparing its performance with theoretical and simulated results.
- Qualitatively seeing the impact of transistor-to-transistor variations.

MATERIALS:

- Laboratory setup, including breadboard
- 1 PNP transistor (e.g., NTE2322)
- 3 large (e.g., 47-μF) capacitors
- Several resistors of varying sizes
- Wires

PART 1: DESIGN AND SIMULATION

Consider the circuit shown in Figure L7.12:

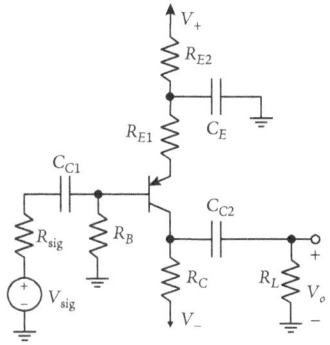

FIGURE L7 12: Common-emitter amplifier circuit, with emitter-degeneration resistor R_{E1}, coupling capacitors, and resistor R_B for DC-biasing purposes. Based on Fig. 7.59 p. 467 S&S.

Design the amplifier to achieve a small-signal gain of at least $A_v = -2.4$ V/V, with $R_{E1} = 2.1$ kΩ. Use supplies of $V_+ = -V_- = 15$ V, $R_{sig} = 50$ Ω, $R_L = 10$ kΩ, and $R_B = 10$ kΩ, and design the circuit to have $I_C = 1$ mA. Obtain the datasheet for the PNP transistor that will be used. In your lab book, perform the following:

DC Operating Point Analysis
- Sketch a DC model of the circuit in your lab book, replacing the three "large-valued" coupling capacitors C_{C1}, C_{C2}, and C_E, by open circuits (for simplicity you may also omit v_{sig}, R_{sig}, and R_L).
- What are the values of I_B and I_E? What is the value of V_B?
- Determine a value of $R_{E1} + R_{E2}$ that produces an emitter-base voltage drop of 0.7 V. What is V_E? What is R_{E2}?
- Are the values of R_{E1} and R_{E2} available in your kit? Can you achieve these values by combining several resistors? Comment.
- *Note*: At this stage we know neither V_{EC} nor R_C.

AC Analysis
- Sketch a small-signal model of the circuit in your lab book, replacing the transistor with its small-signal model (V_A is large, so you may ignore r_o), replacing the capacitors with short circuits (what happens to R_{E2}?), and replacing V_- with an AC ground. What happens to V_+? Label the base of the transistor as v_i, i.e., the small-signal voltage at the input. What are the values of g_m and r_π?
- What is the ratio of v_i/v_{sig}? Can you approximate it?
- Derive an expression for $A_v = v_o/v_i$. What is the value of R_C that produces a small-signal voltage gain of *at least* $A_v = -2.4$ V/V? Is the value you calculate for R_C available in your kit? Can you achieve this value by combining several resistors? Comment.
- What is the DC voltage at the collector? Does this satisfy the assumption that the transistor should be operating in the active region? Explain.

Simulation
- Simulate the performance of your circuit. Use capacitor values $C_{C1} = C_{C2} = C_E = 47$ μF and the values of R_{E1}, R_{E2}, and R_C based on your preceding calculations. Use a 10-mV$_{pk-pk}$, 1-kHz sinusoid with no DC component applied at v_{sig}.
- From your simulation, report the DC values of V_{EB}, V_{EC}, I_B, I_C, and I_E. How closely do they match your calculations?
- From your simulation, report A_v. How closely does it match your calculations?

PART 2: PROTOTYPING

- Assemble the circuit onto your breadboard using the specified component values and those calculated earlier. Note that R_{sig} represents the output resistance of the function generator, and therefore you should *not* include it in your circuit.

PART 3: MEASUREMENTS

- *DC bias point measurement*: Using a digital multimeter, measure the DC voltages of your circuit at the base (V_B), emitter (V_E), and collector (V_C) of your transistor.
- *AC measurement*: Using a function generator, apply a 10-mV$_{pk-pk}$, 1-kHz sinusoid with no DC component to your circuit. (*Note*: Some function generators only allow inputs as small as 50 mV$_{pk-pk}$. If this is the case, use that value instead.)
- Using an oscilloscope, generate plots of v_o and v_i vs. t.
- *Further exploration 1*: What happens to the shape of the output signal as you increase the amplitude of the input signal, e.g., to 1 V$_{pk-pk}$? At what input amplitude do you begin to see significant distortion?
- *Further exploration 2*: What happens if you decrease R_{E1} to 500 Ω but keep $R_{E1} + R_{E2}$ constant?
- Using a digital multimeter, measure all resistors to three significant digits.

PART 4: POST-MEASUREMENT EXERCISE

- Calculate the values of V_{EB} and V_{EC} that you obtained in the lab. How do they compare to your pre-lab calculations? Explain any discrepancies.
- Based on the measured values of V_B, V_C, and V_E and your measured resistor values, what are the real values of all currents based on your lab measurements? How do they compare to your pre-lab calculations? Based on the measured values of currents, what is the actual value of β for your transistor?
- What is the measured value of A_v? How does it compare to your pre-lab calculations? Explain any discrepancies.
- *Hint*: The single biggest source of variations from your pre-lab simulation results will be due to variations in β.

PART 5 [OPTIONAL]: EXTRA EXPLORATION

- In your circuit, switch R_{E1} and R_{E2}. Measure your new DC bias point and small-signal voltage gain. What has changed? What remains the same? Can you explain this?

NPN Common-Base Amplifier
(See Section 7.5.4, p. 469 of Sedra/Smith)

OBJECTIVES:

To study an NPN-based common-base (CB) amplifier by:

- Completing the DC and small-signal analysis based on its theoretical behavior.
- Simulating it to compare the results with the paper analysis.
- Implementing it in an experimental setting, taking measurements, and comparing its performance with theoretical and simulated results.
- Qualitatively seeing the impact of transistor-to-transistor variations.

MATERIALS:

- Laboratory setup, including breadboard
- 1 NPN-type bipolar transistor (e.g., NTE2321)
- 2 large (e.g., 47-μF) capacitors
- Several resistors of varying sizes
- Wires

PART 1: DESIGN AND SIMULATION

Consider the circuit shown in Figure L7.13:

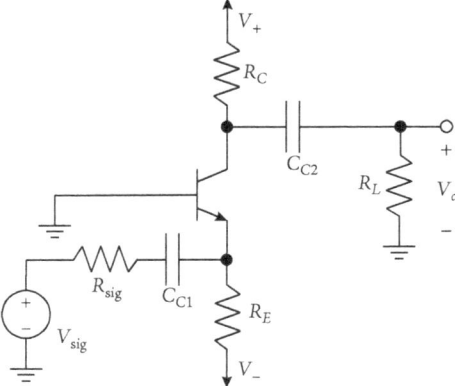

FIGURE L7.13: Common-base amplifier circuit, with coupling capacitors for DC-biasing purposes. Based on Fig. 7.60 p. 470 S&S.

Design the amplifier to achieve a small-signal gain of at least $A_v > 50$ V/V. Use supplies of $V_+ = -V_- = 15$ V, $R_{sig} = 50\ \Omega$, and $R_L = 10$ kΩ, and design the circuit to have $I_C = 1$ mA. Although there will be variations from transistor to transistor, you may assume a value of β of 100 in your calculations. Obtain the datasheet for the NPN transistor that will be used. In your lab book, perform the following.

DC Operating Point Analysis

- Sketch a DC model of the circuit in your lab book, replacing the large-valued coupling capacitors, C_{C1} and C_{C2}, by open circuits (for simplicity you may also omit v_{sig}, R_{sig}, and R_L).
- What are the values of I_B and I_E?
- Determine a value of R_E that produces a base-emitter voltage drop of 0.7 V.
- Is the value of R_E available in your kit? Can you achieve this value by combining several resistors? Comment.
- *Note*: At this stage we know neither V_{CE} nor R_C.

AC Analysis

- Sketch a small-signal model of the circuit in your lab book, replacing the transistor with its small-signal model (ignore r_o), replacing the capacitors with short circuits, and replacing V_+ and V_- with an AC ground. Label the emitter of the transistor as v_i, i.e., the small-signal voltage at the input. What are the values of g_m and r_e?
- What is the ratio of v_i/v_{sig}? Can you approximate it?
- Derive an expression for $A_v = v_o/v_i$. What is the value of R_C that produces a small-signal voltage gain of *at least* $A_v > 50$ V/V? Is the value you calculate for R_C available in your kit? Can you achieve this value by combining several resistors? Comment.
- What is the DC voltage at the collector? Does this satisfy the assumption that the transistor should be operating in the active region? Explain.

Simulation

- Simulate the performance of your circuit. Use capacitor values $C_{C1} = C_{C2} = 47\ \mu$F and the values of R_E and R_C based on your preceding calculations. Use a 10-mV$_{pk-pk}$, 1-kHz sinusoid with no DC component applied at v_{sig}.
- From your simulation, report the DC values of V_{BE}, V_{CE}, I_B, I_C, and I_E. How closely do they match your calculations?
- From your simulation, report A_v. How closely does it match your calculations?

PART 2: PROTOTYPING

- Assemble the circuit onto your breadboard using the specified component values and those just calculated. Note that R_{sig} represents the output resistance of the function generator, and therefore you should *not* include it in your circuit.

PART 3: MEASUREMENTS

- *DC bias point measurement*: Using a digital multimeter, measure the DC voltages of your circuit at the base (V_B), emitter (V_E), and collector (V_C) of your transistor.
- *AC measurement*: Using a function generator, apply a 10-mV$_{pk-pk}$, 1-kHz sinusoid with no DC component to your circuit. (*Note*: Some function generators only allow inputs as small as 50 mV$_{pk-pk}$. If this is the case, use that value instead, but you may expect some distortion in the output waveform.)
- Using an oscilloscope, generate plots of v_i and v_i vs. t.
- *Further exploration*: What happens to the shape of the output signal as you increase the amplitude of the input signal, e.g., to 1 V$_{pk-pk}$? At what input amplitude do you begin to see significant distortion?
- Using a digital multimeter, measure all resistors to three significant digits.

PART 4: POST-MEASUREMENT EXERCISE

- Calculate the values of V_{BE} and V_{CE} that you obtained in the lab. How do they compare to your pre-lab calculations? Explain any discrepancies.
- Based on the measured values of V_C and V_E and your measured resistor values, what are the real values of all currents based on your lab measurements? How do they compare to your pre-lab calculations? Based on the measured values of currents, what is the actual value of β for your transistor?
- What is the measured value of A_v? How does it compare to your pre-lab calculations? Explain any discrepancies.
- *Hint*: The single biggest source of variations from your pre-lab simulation results will be due to variations in β.

PART 5 [OPTIONAL]: EXTRA EXPLORATION

- Using the function generator, add a 1-V DC component to the input v_{sig} and repeat the measurements. Do you still get the same DC operating point and voltage gain A_v? Why or why not?

PNP Common-Base Amplifier
(See Section 7.5.4, p. 469 of Sedra/Smith)

OBJECTIVES:

To study a PNP-based common-base (CB) amplifier by:

* Completing the DC and small-signal analysis based on its theoretical behavior.
* Simulating it to compare the results with the paper analysis.
* Implementing it in an experimental setting, taking measurements, and comparing its performance with theoretical and simulated results.
* Qualitatively seeing the impact of transistor-to-transistor variations.

MATERIALS:

* Laboratory setup, including breadboard
* 1 PNP-type bipolar transistor (e.g., NTE2322)
* 2 large (e.g., 47-μF) capacitors
* Several resistors of varying sizes
* Wires

PART 1: DESIGN AND SIMULATION

Consider the circuit shown in Figure L7.14:

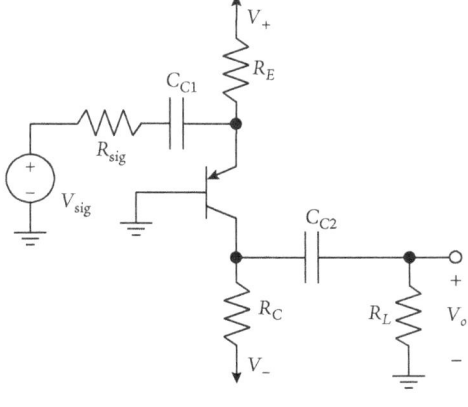

FIGURE L7.14: Common-base amplifier circuit, with coupling capacitors for DC-biasing purposes. Based on Fig. 7.60 p. 470 S&S.

Design the amplifier to achieve a small-signal gain of at least $A_v > 50$ V/V. Use supplies of $V_+ = -V_- = 15$ V, $R_{sig} = 50$ Ω, and $R_L = 10$ kΩ, and design the circuit to have $I_C = 1$ mA. Although there will be variations from transistor to transistor, you may assume a value of β of 100 in your calculations. Obtain the datasheet for the PNP transistor that will be used. In your lab book, perform the following.

DC Operating Point Analysis

- Sketch a DC model of the circuit in your lab book, replacing the large-valued coupling capacitors C_{C1} and C_{C2} by open circuits (for simplicity you may also omit v_{sig}, R_{sig}, and R_L).
- What are the values of I_B and I_E?
- Determine a value of R_E that produces an emitter-base voltage drop of 0.7 V.
- Is the value of R_E available in your kit? Can you achieve this value by combining several resistors? Comment.
- *Note*: At this stage we know neither V_{EC} nor R_C.

AC Analysis

- Sketch a small-signal model of the circuit in your lab book, replacing the transistor with its small-signal model (ignore r_o), replacing the capacitors with short circuits, and replacing V_+ and V_- with an AC ground. Label the emitter of the transistor as v_i, i.e., the small-signal voltage at the input. What are the values of g_m and r_e?
- What is the ratio of v_i/v_{sig}? Can you approximate it?
- Derive an expression for $A_v = v_o/v_i$. What is the value of R_C that produces a small-signal voltage gain of *at least* $A_v > 50$ V/V? Is the value you calculate for R_C available in your kit? Can you achieve this value by combining several resistors? Comment.
- What is the DC voltage at the collector? Does this satisfy the assumption that the transistor should be operating in the active region? Explain.

Simulation

- Simulate the performance of your circuit. Use capacitor values $C_{C1} = C_{C2} = 47$ μF and the values of R_E and R_C based on your preceding calculations. Use a 10-mV$_{pk-pk}$, 1-kHz sinusoid with no DC component applied at v_{sig}.
- From your simulation, report the DC values of V_{EB}, V_{EC}, I_B, I_C, and I_E. How closely do they match your calculations?
- From your simulation, report A_v. How closely does it match your calculations?

PART 2: PROTOTYPING

- Assemble the circuit onto your breadboard using the specified component values and those just calculated. Note that R_{sig} represents the output resistance of the function generator, and therefore you should *not* include it in your circuit.

PART 3: MEASUREMENTS

- *DC bias point measurement*: Using a digital multimeter, measure the DC voltages of your circuit at the base (V_B), emitter (V_E), and collector (V_C) of your transistor.
- *AC measurement*: Using a function generator, apply a 10-mV$_{pk-pk}$, 1-kHz sinusoid with no DC component to your circuit. (*Note*: Some function generators only allow inputs as small as 50 mV$_{pk-pk}$. If this is the case, use that value instead, but you may expect some distortion in the output waveform.)
- Using an oscilloscope, generate plots of v_o and v_i vs. t.
- *Further exploration*: What happens to the shape of the output signal as you increase the amplitude of the input signal, e.g., to 1 V$_{pk-pk}$? At what input amplitude do you begin to see significant distortion?
- Using a digital multimeter, measure all resistors to three significant digits.

PART 4: POST-MEASUREMENT EXERCISE

- Calculate the values of V_{EB} and V_{EC} that you obtained in the lab. How do they compare to your pre-lab calculations? Explain any discrepancies.
- Based on the measured values of V_C and V_E and your measured resistor values, what are the real values of all currents based on your lab measurements? How do they compare to your pre-lab calculations? Based on the measured values of currents, what is the actual value of β for your transistor?
- What is the measured value of A_v? How does it compare to your pre-lab calculations? Explain any discrepancies.
- *Hint*: The single biggest source of variations from your pre-lab simulation results will be due to variations in β

PART 5 [OPTIONAL]: EXTRA EXPLORATION

- Using the function generator, add a 1-V DC component to the input v_{sig} and repeat the measurements. Do you still get the same DC operating point and voltage gain A_v? Why or why not?

NPN Emitter Follower
(See Section 7.5.5, p. 471 of Sedra/Smith)

OBJECTIVES:

To study an NPN-based emitter follower by:

- Completing the DC and small-signal analysis based on its theoretical behavior.
- Simulating it to compare the results with the paper analysis.
- Implementing it in an experimental setting, taking measurements, and comparing its performance with theoretical and simulated results.
- Qualitatively seeing the impact of transistor-to-transistor variations.

MATERIALS:

- Laboratory setup, including breadboard
- 1 NPN transistor (e.g., NTE2321)
- 3 large (e.g., 47-μF) capacitors
- Several resistors of varying sizes
- Wires

PART 1: DESIGN AND SIMULATION

Consider the circuit shown in Figure L7.15:

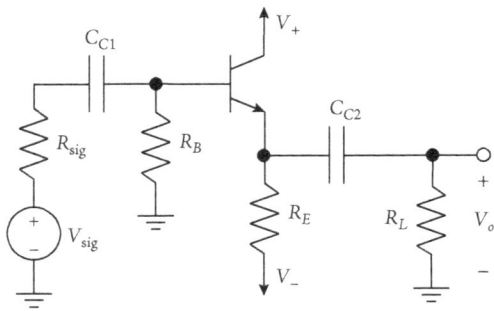

FIGURE L7.15: Emitter-follower circuit, with coupling capacitors, and resistor R_B for DC-biasing purposes. Based on Fig. 7.61 p. 471 S&S.

Design the amplifier such that $I_C = 1$ mA and $A_v = 0.95$ V/V. Use supplies of $V_+ = -V_- = 15$ V, $R_{sig} = 50$ Ω, and $R_B = 10$ kΩ. What is the minimum value of R_L that satisfies the design requirements? Obtain the datasheet for the NPN transistor that will be used. In your lab book, perform the following.

DC Operating Point Analysis

- Sketch a DC model of the circuit in your lab book, replacing the large-valued coupling capacitors, C_{C1} and C_{C2}, by open circuits (for simplicity you may also omit v_{sig}, R_{sig}, and R_L). What is I_E? What are V_B and V_E?
- Based on the required value of I_C, what is R_E?

AC Analysis

- Sketch a small-signal model of the circuit in your lab book, replacing the transistor with its small-signal model (try a T model, ignoring r_o), replacing the capacitors with short circuits, and replacing V_+ and V_- with an AC ground. Label the base of the transistor as v_i, i.e., the small-signal voltage at the input.
- What is the ratio of v_i/v_{sig}? How would you approximate it in further calculations?
- Derive an expression for $A_v = v_o/v_i$.
- What is the minimum value of R_L that satisfies the design requirements?
- What is the value of g_m? What is r_e? What is A_v?
- Calculate the output resistance of your amplifier.

Simulation

- Simulate the performance of your circuit. Use capacitor values $C_{C1} = C_{C2} = 47$ μF and the value of R_E based on your preceding calculations. Use a 10-mV$_{pk-pk}$, 1-kHz sinusoid with no DC component applied at v_{sig}.
- From your simulation, report the DC values of V_{BE}, V_{CE}, and I_C. How closely do they match your calculations? (Remember: The simulator has its own, more complex model of the real transistor, so there should be some small variations.)
- From your simulation, report A_v. How closely does it match your calculations?

PART 2: PROTOTYPING

- Assemble the circuit onto your breadboard using the specified component values and those just calculated. Note that R_{sig} represents the output resistance of the function generator, and therefore you should *not* include it in your circuit.

PART 3: MEASUREMENTS

- *DC bias point measurement*: Using a digital multimeter, measure the DC voltages of your circuit at the base (V_B) and emitter (V_E) of your transistor.

- *AC measurement*: Using a function generator, apply a 10-mV$_\text{pk-pk}$, 1-kHz sinusoid with no DC component to your circuit. (*Note*: Some function generators only allow inputs as small as 50 mV$_\text{pk-pk}$. If this is the case, use that value instead.)
- Using an oscilloscope, generate plots of v_o and v_i vs. t.
- Using a digital multimeter, measure all resistors to three significant digits.

PART 4: POST-MEASUREMENT EXERCISE

- Calculate the values of V_{BE} and V_{CE} that you obtained in the lab. How do they compare to your pre-lab calculations? Explain any discrepancies.
- Based on the measured values of V_B and V_E and your measured resistor values, what is the real value of I_C based on your lab measurements?
- What is the measured value of A_v? How does it compare to your pre-lab calculations? Explain any discrepancies.
- What would happen if you used the function generator with 50-Ω output resistance to drive your load resistor directly? What gain would you get? What would happen if the output resistance of the function generation was changed from 50 Ω to 5 kΩ? What do you conclude? Recall the value of output resistance you calculated earlier.
- *Hint*: The single biggest source of variations from your pre-lab simulation results will be due to variations in β.

PART 5 [OPTIONAL]: EXTRA EXPLORATION

- Add a 500-Ω resistor between the function generator output and capacitor C_{C1}. How does the gain of your circuit change? Can you explain this?

PNP Emitter Follower
(See Section 7.5.5, p. 469 of Sedra/Smith)

OBJECTIVES:

To study a PNP-based emitter follower by

- Completing the DC and small-signal analysis based on its theoretical behavior.
- Simulating it to compare the results with the paper analysis.
- Implementing it in an experimental setting, taking measurements, and comparing its performance with theoretical and simulated results.
- Qualitatively seeing the impact of transistor-to-transistor variations.

MATERIALS:

- Laboratory setup, including breadboard
- 1 PNP transistor (e.g., NTE2322)
- 3 large (e.g., 47-μF) capacitors
- Several resistors of varying sizes
- Wires

PART 1: DESIGN AND SIMULATION

Consider the circuit shown in Figure L7.16:

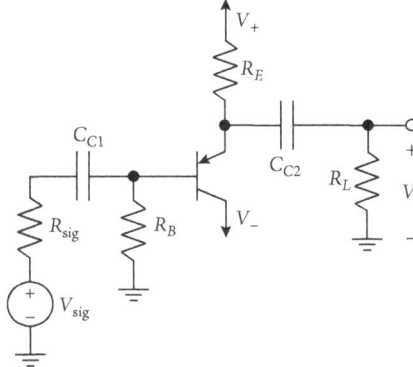

FIGURE L7.16: Emitter follower circuit, with coupling capacitors, and resistor R_B for DC-biasing purposes. Based on Fig. 7.61 p. 471 S&S.

Design the amplifier such that $I_C = 1$ mA and $A_v = 0.95$ V/V. Use supplies of $V_+ = -V_- = 15$ V, $R_{sig} = 50$ Ω, and $R_B = 10$ kΩ. What is the minimum value of R_L that satisfies the design requirements? Obtain the datasheet for the PNP transistor that will be used. In your lab book, perform the following.

DC Operating Point Analysis

- Sketch a DC model of the circuit in your lab book, replacing the large-valued coupling capacitors, C_{C1} and C_{C2}, by open circuits (for simplicity you may also omit v_{sig}, R_{sig}, and R_L). What is I_B? What are V_B and V_E?
- Based on the required value of I_C, what is R_E?

AC Analysis

- Sketch a small-signal model of the circuit in your lab book, replacing the transistor with its small-signal model (try a T model, ignoring r_o), replacing the capacitors with short circuits, and replacing V_+ and V_- with an AC ground. Label the base of the transistor as v_i, i.e., the small-signal voltage at the input.
- What is the ratio of v_i/v_{sig}? How would you approximate it in further calculations?
- Derive an expression for $A_v = v_o/v_i$.
- What is the minimum value of R_L that satisfies the design requirements?
- What is the value of g_m? What is r_e? What is A_v?
- Calculate the output resistance of your amplifier.

Simulation

- Simulate the performance of your circuit. Use capacitor values $C_{C1} = C_{C2} = 47$ μF and the value of R_E based on your preceding calculations. Use a 10-mV$_{pk-pk}$, 1-kHz sinusoid with no DC component applied at v_{sig}.
- From your simulation, report the DC values of V_{EB}, V_{EC}, and I_C. How closely do they match your calculations? (Remember: The simulator has its own, more complex model of the real transistor, so there should be some small variations.)
- From your simulation, report A_v. How closely does it match your calculations?

PART 2: PROTOTYPING

- Assemble the circuit onto your breadboard using the specified component values and those just calculated. Note that R_{sig} represents the output resistance of the function generator, and therefore you should *not* include it in your circuit.

PART 3: MEASUREMENTS

- *DC bias point measurement*: Using a digital multimeter, measure the DC voltages of your circuit at the base (V_B) and emitter (V_E) of your transistor.

- *AC measurement*: Using a function generator, apply a 10-mV$_{pk-pk}$, 1-kHz sinusoid with no DC component to your circuit. (*Note*: Some function generators only allow inputs as small as 50 mV$_{pk-pk}$. If this is the case, use that value instead.)
- Using an oscilloscope, generate plots of v_o and v_i vs. t.
- Using a digital multimeter, measure all resistors to three significant digits.

PART 4: POST-MEASUREMENT EXERCISE

- Calculate the values of V_{EB} and V_{EC} that you obtained in the lab. How do they compare to your pre-lab calculations? Explain any discrepancies.
- Based on the measured values of V_B and V_E and your measured resistor values, what is the real value of I_C based on your lab measurements?
- What is the measured value of A_v? How does it compare to your pre-lab calculations? Explain any discrepancies.
- What would happen if you used the function generator with 50-Ω output resistance to drive your load resistor directly? What gain would you get? What would happen if the output resistance of the function generation was changed from 50 Ω to 5 kΩ? What do you conclude? Recall the value of output resistance you calculated earlier.
- *Hint*: The single biggest source of variations from your pre-lab simulation results will be due to variations in β.

PART 5 [OPTIONAL]: EXTRA EXPLORATION

- Add a 500-Ω resistor between the function generator output and capacitor C_{C1}. How does the gain of your circuit change? Can you explain this?

NMOS vs. NPN: Common-Source/ Common-Emitter Amplifier Comparison

OBJECTIVES:

To compare MOS- and BJT-based amplifiers by:

- Building and characterizing an NMOS-based common-source (CS) amplifier.
- Replacing the NMOS transistor with an NPN transistor and recharacterizing the new common-emitter (CE) circuit.

MATERIALS:

- Laboratory setup, including breadboard
- 1 enhancement-type NMOS transistor (e.g., MC14007)
- 1 NPN transistor (e.g., NTE2321)
- 3 large (e.g., 47-μF) capacitors
- Several resistors of varying sizes
- Wires

Consider the two circuits shown in Figure L7.17. Note the structural similarity between the two circuits.

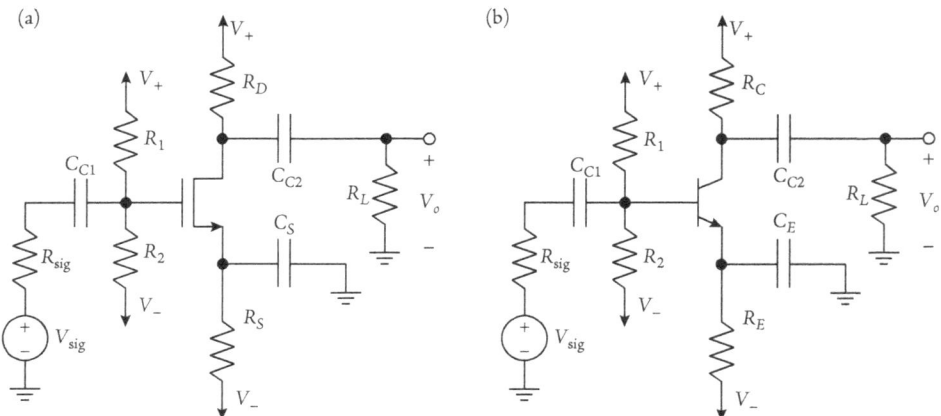

FIGURE L7.17: (a) NMOS common-source amplifier, and (b) NPN common-emitter amplifier.

PART 1: CS DESIGN AND MEASUREMENT

Hand calculations

Design the CS amplifier in Fig. L7.17(a) to achieve a small-signal gain of at least $A_v = -5$ V/V. Use supplies of $V_+ = -V_- = 15$ V, $R_{sig} = 50\ \Omega$, $R_L = 10$ kΩ, and $R_1 \| R_2 = 10$ kΩ, and design the circuit to have $I_D = 1$ mA and a DC voltage at the gate $V_G = 0$ V. Use $C_{C1} = C_{C2} = C_S = 47\ \mu$F. What is the expected DC voltage at the source of the NMOS?

Simulations

Simulate the performance of your circuit. Use a 10-mV$_{pk-pk}$, 1-kHz sinusoid with no DC component applied at v_{sig}. Report the DC value of V_S. What is A_v?

Prototyping

Assemble the circuit onto your breadboard using the specified component values and those just calculated. Once more, R_{sig} represents the output resistance of the function generator, and therefore you should *not* include it in your circuit.

Measurements

Using a digital multimeter, measure the DC voltages of your circuit at the gate (V_G), source (V_S), and drain (V_D) of your transistor. Then, using a function generator, apply a 50-mV$_{pk-pk}$, 1-kHz sinusoid with no DC component to your circuit. Using an oscilloscope, generate plots of V_0 and v_i vs. t. What is A_v?

PART 2: CE DESIGN AND MEASUREMENT

Hand calculations

Consider the CE amplifier in Fig. L7.17(b). The main difference from the CS amplifier is that we have replaced the NMOS transistor by an NPN transistor (and changed some of the notation!).

Setting $R_E = R_S$, $R_C = R_D$, and $C_E = C_S$, and keeping $V_E = V_S$ (use your measured value) and $I_D = I_E$, what DC voltage V_B do you require at the base of the NPN transistor? (*Hint*: An active NPN transistor has a base-emitter voltage of approximately 0.7 V.) What new values of R_1 and R_2 do you need to use to achieve this value of V_B while keeping $R_1 \| R_2 = 10$ kΩ? Since $I_B \neq 0$, you will need to derive the Thévenin equivalent circuit.

Simulation

Simulate the common-emitter circuit using the earlier values (including the new values of R_1 and R_2). What is the voltage gain of the new circuit? What are V_B, V_C, and V_E? How do they compare to V_G, V_D, and V_S in the CS circuit?

Prototyping

Assemble the new circuit onto your breadboard using the specified component values and those just calculated. You should be able just to replace the NMOS with an NPN transistor, and replace R_1 and R_2 with their new values.

Measurements

Using a digital multimeter, measure the DC voltages of your circuit at the base (V_B), emitter (V_E), and collector (V_C) of your transistor. Then, using a function generator, apply a 50-mV$_{pk-pk}$, 1-kHz sinusoid with no DC component to your circuit. Using an oscilloscope, generate plots of v_o and v_i vs. t. What is A_v?

PART 3: COMPARISON

List all the similarities and differences you can think of between the two amplifiers.

CONCLUDING REMARKS

The authors wish you well in your continued studies!

And remember . . . it will all work out in the end. If it hasn't worked out yet, then it clearly isn't the end!